CISM COURSES AND LECTURES

Series Editors:

The Rectors of CISM
Sandor Kaliszky - Budapest
Mahir Sayir - Zurich
Wilhelm Schneider - Wien

The Secretary General of CISM
Giovanni Bianchi - Milan

Executive Editor
Carlo Tasso - Udine

The series presents lecture notes, monographs, edited works and proceedings in the field of Mechanics, Engineering, Computer Science and Applied Mathematics.
Purpose of the series is to make known in the international scientific and technical community results obtained in some of the activities organized by CISM, the International Centre for Mechanical Sciences.

INTERNATIONAL CENTRE FOR MECHANICAL SCIENCES

COURSES AND LECTURES - No. 351

SHAPE MEMORY ALLOYS

M. FREMOND
LAB. DES MATERIAUX ET STRUCTURES DU GENIE CIVIL
S. MIYAZAKI
UNIVERSITY OF TSUKUBA

Springer-Verlag Wien GmbH

Le spese di stampa di questo volume sono in parte coperte da contributi del Consiglio Nazionale delle Ricerche.

This volume contains 99 illustrations

Originally published by Springer-Verlag Wien New York in 1996

In order to make this volume available as economically and as rapidly as possible the authors' typescripts have been reproduced in their original forms. This method unfortunately has its typographical limitations but it is hoped that they in no way distract the reader.

ISBN 978-3-211-82804-5 ISBN 978-3-7091-4348-3 (eBook)
DOI 10.1007/978-3-7091-4348-3

PREFACE

The thermal and mechanical properties of shape memory alloys are puzzling: thermal actions induce large deformations. In some sense thermal actions are equivalent to mechanical actions. This equivalence is the basic tool for many practical applications. The complete description of the shape memory alloys involves physics, metallurgy, crystallography, mechanics and even mathematics to deal with the predictive theories describing their behaviour.

This book consisting of two chapters is devoted to the crystallographic and thermomechanical properties and to the macroscopic modelling of the shape memory alloys. It results from part of the lectures given by professors Fremond, James, Miyazaki and Müller during a session of the Centre International des Sciences Mécaniques from October 4 to October 8, 1993.

The first chapter describes the thermomechanical macroscopic theory. Shape memory alloys are mixtures of many phases: martensites and austenite. The composition of the mixtures vary: the martensites and the austenite which have different mechanical properties transform into one another. The phase changes can be produced either by thermal actions or by mechanical actions. The striking and well known properties of shape memory alloys results from these links between mechanical and thermal actions.
The macroscopic modelling is based on thermodynamics involving internal quantities. It gives a macroscopic theory which can be used for engineering purposes, for example to describe the evolution of structures made of shape memory alloys. Some of the internal quantities which are chosen, the phase volume fractions for instance, are submitted to constraints (for instance their actual value is between 0 and 1). It is shown that most of the macroscopic properties of shape memory alloys result from a careful treatment of these constraints. The thermomechanical modelling uses continuum thermodynamics and some elementary notions of convex analysis as basic tools.

The second chapter describes the experimental works on the crystallographic and thermomechanical properties. Recent development of shape memory alloys is reviewed, emphasis being placed on the Ti-Ni, Cu-based and ferrous alloys which are considered as practical materials for applications among many shape memory alloys. Crystal structures of the parent and martensitic phases are described, and the crystallography of the martensitic transformations is also briefly explained. The origin of the shape memory effect and the shape memory mechanisms are discussed on the basis of the crystal structure and the crystallography of the martensitic transformations. Since an applied stress also induces the martensitic transformations, successive stages of the stress-induced martensitic transformations are reviewed briefly in Cu-based and Ti-Ni alloys, which show martensite-to-martensite transformations upon loading. Then, the transformation and mechanical characteristics of the shape memory alloys are reviewed in detail; i.e. phase diagrams, transformation temperatures, transformation process, stress-induced transformation, aging effects, cycling effects, fracture, fatigue, grain refinement, two-way shape memory effect, and so on. Recent development of sputter-deposited Ti-Ni thin films is also introduced.

The authors wish that this book would be a good guide and introduction for the readers to the study of shape memory alloys.

M. Fremond
S. Miyazaki

CONTENTS

SHAPE MEMORY ALLOY
A THERMOMECHANICAL MACROSCOPIC THEORY

M. Frémond
Lab. des Matériaux et Structures du Génie Civil, Champs sur Marne, France

1. Introduction.

Shape memory alloys are mixtures of many martensites and of austenite. The composition of the mixture varies : the matensites and the austenite transform into one another. These phase changes can be produced either by thermal actions or by mechanical actions. The striking and well known properties of shape memory alloys results from these links between mechanical and thermal actions [4], [15], [26].

Shape memory alloys can be studied at the microscopic level by describing the microstructures of the constitutive crystals, [7], [17], [28]. They can also be studied by using statistical thermodynamics of a lattice of particles [3] [23].

Thermodynamics involving internal quantities is an other tool to study shape memory alloys at the macroscopic level, [1], [2], [5], [6], [9], [18], [19], [22] [25], [27], [29]. It is the one we have chosen [10], [11]. It gives a macroscopic theory which can be used for engineering purposes, for example to describe the evolution of structures made of shape memory alloys. The internal quantities we choose, the phase volume fractions, are submitted to constraints (for instance their actual value is between 0 and 1). We show that most of the properties of shape memory alloys result from a careful treatment of these constraints [10], [12].

The first paragraphs 1 to 8 describe the thermodynamical quantities, the free energy and the pseudo-potential of dissipation. They give also the basic tools for macroscopic modelling. The paragraph 9 is devoted to the macroscopic description of shape memory alloys.

Basic definitions and properties of convex analysis are given in the appendix 12.

2. Description of a material. The state quantities.

The state quantities are the basic quantities which describe the equilibrium and the evolution of a material. Their choice depends on the sophistication of the model we are searching for. Thus their choice depends on the scientist or the engineer concerned.

When the state quantities are constant with respect to the time, we say that the material is at an equilibrium. Thus the notion of equilibrium is subjective : it depends on the sophistication of the description.

The set of the state quantities is denoted by E. It usually contains quantities describing the deformations and the temperature. The other quantities of E are often called internal quantities.

3. Principle of Virtual Pøwer without micoscopic velocities.

This is the classical situation. Let $\mathcal{V}$ be the linear space of the macroscopic virtual velocities, Ω be the domain of $\mathbf{R}^3$ occupied by the structure we consider at the time t. The principle of virtual power [14] is

$$(1), \quad \forall D \subset \Omega, \forall \mathbf{V} \in \mathcal{V}, A(D,\mathbf{V}) = P_i(D,\mathbf{V}) + P_e(D,\mathbf{V}),$$

where D is a subdomain of Ω. The virtual power of the acceleration forces is

$$A(D,\mathbf{V}) = \int_D \rho\gamma\mathbf{V}d\Omega,$$

where γ is the acceleration and ρ the density. The virtual power of the internal forces is

$$P_i(D,\mathbf{V}) = -\int_D \sigma:D(\mathbf{V})d\Omega,$$

where $D(V) = (D_{ij}(V) = \frac{1}{2}(V_{i,j}+V_{j,i}))$ are the strain rates and $\sigma = (\sigma_{ij})$ the stresses. The virtual power of the external forces is the sum of the power of the at a distance forces,

$$\int_D \mathbf{f}.\mathbf{V}d\Omega,$$

where f is the volumetric external force, and of the power of the contact forces

$$\int_{\partial D} \mathbf{T}.\mathbf{V}d\Gamma,$$

where T is the contact external force. The virtual power of the external forces is

$$P_e(D,\mathbf{V}) = \int_D \mathbf{f}.\mathbf{V}d\Omega + \int_{\partial D} \mathbf{T}.\mathbf{V}d\Gamma.$$

It is classical [14] to get the equations of the movement from the principle (1) ,

(2), $\rho\gamma = \mathrm{div}\sigma + \mathbf{f}$, in D,

(3), $\sigma.\mathbf{N} = \mathbf{T}$, in ∂D,

where N is the outwards normal unit vector to D.

4. Principle of Virtual Power with microscopic velocities.

When the state quantities include internal quantities, the evolution of those quantities can result from microscopic movements. We think that the power of the microscopic movements can be taken into account in the power of the internal forces [13]. Let β be an internal quantity, for instance a volumetric proportion of austenite in a shape memory alloy, the volumetric proportion of a constituent in a mixture, the damage in a piece of concrete [13], the intensity of adhesion between two pieces [16], [32], the volume fraction of unfrozen water in a soil in winter... The only macroscopic quantity which is related to the micoscopic movements or velocities is $\frac{d\beta}{dt}$ which describes their macroscopic effects i.e. the evolution of β. Thus we choose as actual power of the internal forces

$$P_i(D,\mathbf{U},\frac{d\beta}{dt}) = -\int_D \sigma:D(\mathbf{U})d\Omega - \int_D \{B\frac{d\beta}{dt} + \mathbf{H}.\mathbf{grad}\frac{d\beta}{dt}\}d\Omega,$$

and as virtual power of the internal forces,

$$\forall(\mathbf{V},c)\in \mathbb{V}\times\mathbb{C}, \qquad P_i(D,\mathbf{V},c) = -\int_D \sigma:D(\mathbf{V})d\Omega - \int_D \{Bc + \mathbf{H}.\mathbf{grad}c\}d\Omega,$$

where $\mathbb{C}$ is the linear space of the virtual microscopic velocities. The elements $(\mathbf{V},c)$ of $\mathbb{V}\times\mathbb{C}$ are function of $\mathbf{x}$, $(\mathbf{V}(\mathbf{x}),c(\mathbf{x}))$.

The gradient of the velocity of β takes into account the influence of the neighbourhood of a point onto this point. The two quantities B and **H** are new internal forces. The force B is a work and **H** is a work flux vector (if β is a volumetric proportion). Their physical meaning will be given by the equations of movement as the physical meaning of the stress tensor is given by (3). We will see that **H** like σ drescribes the effects of the neighbourhood of a point onto this point. The power of the internal forces has to satisfy the virtual power axiom [14]:

the power of the internal forces is zero for any rigid body movement.

A rigid body movement is such that the distance of two material points remains constant. It results that D(U) = 0 for a rigid body movement with macroscopic velocity U. Because the distance of two points remains constant in a rigid body velocity there is no microscopic movement and the value of β remains constant. It results that $\frac{d\beta}{dt} = 0$ and $P_i(D,\mathbf{U},\frac{d\beta}{dt}) = 0$. The virtual power is then satisfied by the power of the internal forces P_i.

It is natural to choose a new power of the external forces $P_e(D,\mathbf{U}\frac{d\beta}{dt})$ depending on $\frac{d\beta}{dt}$. It is the sum of

$$\int_D \mathbf{f}.\mathbf{U}d\Omega + \int_D A\frac{d\beta}{dt}d\Omega,$$

the power of the at a distance external actions, and of the power of the external contact forces

$$\int_{\partial D} \mathbf{T}.\mathbf{U}d\Gamma + \int_{\partial D} a\frac{d\beta}{dt}d\Gamma,$$

where A is the volumic work provided from the exterior and a the surfacic work provided by contact to D. Thus the new power of the external forces we choose is

$$\forall(\mathbf{V},c)\in \mathbb{V}\mathrm{x}\mathbb{C}, P_e(D,\mathbf{V},c) = \int_D \mathbf{f}.\mathbf{V}d\Omega + \int_D Acd\Omega + \int_{\partial D} \mathbf{T}.\mathbf{V}d\Gamma + \int_{\partial D} acd\Gamma.$$

We decide not to change the power of the acceleration forces. The principle of virtual power becomes [12],

(4), $\quad \forall D\subset\Omega, \forall(\mathbf{V},c)\in \mathbb{V}\mathrm{x}\mathbb{C}, \qquad A(D,\mathbf{V},c) = P_i(D,\mathbf{V},c) + P_e(D,\mathbf{V},c).$

By letting c = 0 in (4), we get the classical equations of movement,

(2), $\quad \rho\gamma = \mathrm{div}\sigma + \mathbf{f}$, in D,

(3), $\quad \sigma.\mathbf{N} = \mathbf{T}$, in ∂D.

By letting $\mathbf{V} = \mathbf{0}$ in (4), an easy computation gives,

(5), $\quad 0 = \mathrm{div}\mathbf{H} - B + A$, in D,

(6), $\quad \mathbf{H}.\mathbf{N} = a$, in ∂D.

The equation (6) gives the physical meaning of **H**. It is a work flux vector : **H.N** is the amount of work provided to the body through the surface with normal **N** (for an analoguous situation think of the heat flux vector).

When the power of the internal forces does not depend on $\mathbf{grad}\frac{d\beta}{dt}$ the principle of virtual power gives

(5 bis), $0 = -B + A$, in D.

Of course this equation can be obtained by letting $\mathbf{H} = \mathbf{0}$ in (5).

5. <u>The energy balance.</u>

The conservation of energy is for any subdomain D,

$$\frac{d}{dt}\int_D e\,dt + \frac{dK}{dt} = P_e(D,\mathbf{U},\frac{d\beta}{dt}) + \int_{\partial D} -\mathbf{q}.\mathbf{N}\,d\Gamma + \int_D r\,dt,$$

where e is the volumic internal energy, K the kinetic energy, **q** the heat flux vector and r the volumic rate of heat production. By using the kinetic energy theorem, i.e. the principle of virtual power with the actual velocities, we get for any D,

$$\frac{d}{dt}\int_D e\,dt = -P_i(D,\mathbf{U},\frac{d\beta}{dt}) + \int_{\partial D} -\mathbf{q}.\mathbf{N}\,d\Gamma + \int_D r\,dt.$$

This equation gives,

(7), $$\frac{de}{dt} + e\,\mathrm{div}\mathbf{U} + \mathrm{div}\mathbf{q} = r + \sigma:D(\mathbf{U}) + B\frac{d\beta}{dt} + \mathbf{H}.\mathbf{grad}\frac{d\beta}{dt}, \text{ in } \Omega,$$

(8), $-\mathbf{q}.\mathbf{N} = \pi$, in $\partial\Omega$,

where π is the rate of heat provided to the stucture Ω by contact actions.

6. <u>The second principle of thermodynamics.</u>

It is for any domain$D$,

$$\frac{d}{dt}\int_D s\,dt \geq \int_{\partial D} \frac{-\mathbf{q}.\mathbf{N}}{T}\,d\Gamma + \int_D \frac{r}{T}\,dt,$$

where s is the volumic entropy and T the temperature. It gives,

(9), $$\frac{ds}{dt} + s\,\mathrm{div}\mathbf{U} + \mathrm{div}\frac{\mathbf{q}}{T} \geq \frac{r}{T}.$$

This equation is the second principle basic relation. It is to be satisfied like the balance equations by any actual evolution. The equation (9) multiplied by the temperature assumed to be positive and the energy balance equation give,

$$(10),\quad \frac{de}{dt}-T\frac{ds}{dt}+(e-Ts)\mathrm{div}\mathbf{U}\le\sigma{:}D(U)+B\frac{d\beta}{dt}+\mathbf{H}.\mathbf{grad}\frac{d\beta}{dt}+\frac{-\mathbf{q}.\mathbf{grad}T}{T},$$

by letting $\Psi = e-Ts$ the volumic free energy, we get

$$(11),\quad \frac{d\Psi}{dt}+s\frac{dT}{dt}\le(\sigma-\Psi 1){:}D(U)+B\frac{d\beta}{dt}+\mathbf{H}.\mathbf{grad}\frac{d\beta}{dt}+\frac{-\mathbf{q}.\mathbf{grad}T}{T},$$

which is the Claudius-Duhem relation (1 is the identiy tensor). This inequality is to be satisfied by any actual evolution.

7. Constitutive laws when there are no constraint on the internal quantities.

Let us assume that the internal quantity β can have any value : it is not submitted to any constraint.

From now on, for the sake of simplicity, we make the small perturbation assumption. The Clausius-Duhem inequality becomes

$$(11),\quad \frac{d\Psi}{dt}+s\frac{dT}{dt}\le\sigma{:}D(U)+B\frac{d\beta}{dt}+\mathbf{H}.\mathbf{grad}\frac{d\beta}{dt}+\frac{-\mathbf{q}.\mathbf{grad}T}{T},$$

because $e\mathrm{div}\mathbf{U}$, $s\mathrm{div}\mathbf{U}$ and $\Psi\mathrm{div}\mathbf{U}$ are negligeable in the small perturbation theory. We assume that the state quantities are the small deformations ε, the internal quantity β, its gradient $\mathbf{grad}\beta$ and the temperature T: $E = (\varepsilon,\beta,\mathbf{grad}\beta,T)$. The free energy depends on E: $\Psi(E)$. We assume it is differentiable and let

$$(12),\quad s=-\frac{\partial\Psi}{\partial T},$$

which is the Helmholtz relation, and define,

$$\sigma^{nd}=\frac{\partial\Psi}{\partial\varepsilon},$$

$$B^{nd}=\frac{\partial\Psi}{\partial\beta},$$

$$(13),\quad \mathbf{H}^{nd}=\frac{\partial\Psi}{\partial(\mathbf{grad}\beta)},$$

The stress σ^{nd} is the non-dissipative stress, B^{nd} and $\mathbf{H}^{nd}$ are is the generalized non-dissipative forces. The Clausius-Duhem relation (11) gives

$$(\sigma-\sigma^{nd}){:}D(U)+(B-B^{nd})\frac{d\beta}{dt}+(\mathbf{H}-\mathbf{H}^{nd}).\mathbf{grad}\frac{d\beta}{dt}+T\mathbf{q}.\mathbf{grad}\frac{1}{T}\ge 0,$$

for any actual evolution of the structure. To achieve the description of the constitutive laws, we assume that there exist four functions σ^d, B^d, $\mathbf{H}^d$, $\mathbf{Q}^d$ depending on x, t, E, $\delta E = \{D(U),\frac{d\beta}{dt},\mathbf{grad}\frac{d\beta}{dt},\mathbf{grad}\frac{1}{T}\}$ and other quantities χ depending on the history of the material such that

$$\forall \mathbf{x},t,\forall E,\ \forall W = (f,c,g,,p) \in S x \mathbf{R} x \mathbf{R}^3 x \mathbf{R}^3,\ \forall \chi,$$

(14), $\sigma^d(\mathbf{x},t,E,W,\chi):f + B^d(\mathbf{x},t,E,W,\chi)c + \mathbf{H}^d(\mathbf{x},t,E,W,\chi).\mathbf{g} + \mathbf{Q}^d(\mathbf{x},t,E,W,\chi).\mathbf{p} \geq 0,$

where S is the set of symetric tensors.

The quantity σ^d is the dissipative stress, B^d and $\mathbf{H}^d$ are the dissipative generalized forces and $\mathbf{Q}^d$ is a heat flux. The constitutive laws we choose are,

(15), $\sigma = \sigma^{nd}(E) + \sigma^d(\mathbf{x},t,E,\delta E,\chi),$

(16), $B = B^{nd}(E) + B^d(\mathbf{x},t,E,\delta E,\chi),$

(17), $\mathbf{H} = \mathbf{H}^{nd}(E) + \mathbf{H}^d(\mathbf{x},t,E,\delta E,\chi),$

(18), $T\mathbf{q} = \mathbf{Q}^d(\mathbf{x},t,E,\delta E,\chi).$

It is very easy to prove that

Theorem 1. If the relation (14) are satisfied, the constitutive laws (15) to (18) are such that the Clausius-Duhem inequality is satisfied.

Proof. Write the relation (14) with actual velocities and use the constitutive laws.

Let us sum up the way we define a material : it is defined by

the state quantities E,

the linear spaces $\mathbb{V}$ and $\mathbb{C}$,

the power of the internal forces $P_i(D,\mathbf{V},c)$,

the free energy Ψ depending on E,

the functions σ^d, B^d, $\mathbf{H}^d$, $\mathbf{Q}^d$ depending on E, δE and other quantities χ.

7.1. The pseudo-potential of dissipation.

A very general and powerful method to define the dissipative forces is to introduce a pseudo-potential of dissipation as defined by Moreau [21]. It is a function Φ of E, W and χ such that Φ is a positive function, convex with respect to W [20] and equal to 0 for W = 0. Let us prove,

Theorem 2. Let there be $\Phi(E,W,\chi)$ such that $\delta E \rightarrow \Phi(E,W,\chi)$ is convex, $\Phi(E,W,\chi)$ is positive and $\Phi(E,0,\chi) = 0$. Let for any W, $A \in \partial\Phi(E,W,\chi)$, then we have,

$$A.W \geq 0,$$

(A.W is the scalar product of A and W and $\partial\Phi(E,W,\chi)$ is the subdifferential set of Φ with respect to W).

Proof. Because $W \to \Phi((E,W,\chi))$ is convex we have for $A \in \partial\Phi((E,W,\chi)$,

$$\Phi(E,0,\chi) \geq \Phi(E,W,\chi) - A.W,$$

which gives

$$A.W \geq \Phi(E,W,\chi) \geq 0 \text{ and } A.W \geq 0.$$

The dissipative forces σ^d, B^d, H^d, Q^d are defined by

$$(\sigma^d(x,t,E,\delta E,\chi), B^d(x,t,E,\delta E,\chi), H^d(x,t,E,\delta E,\chi), Q^d(x,t,E,\delta E,\chi)) \in \partial\Phi((E,\delta E,\chi).$$

It results from the theorem 2 that the inequality (14) is satisfied. In the sequel we will always define the dissipative forces with a pseudo-potential of dissipation.

8. Constitutive laws when there are constraints on the internal quantities.

Let us assume that β is a volume fraction or a damage quantity. Because in the following paragraphs the β's will be volume fractions, we choose β to be a damage quantity [13]. The damaged quantity can be defined as the quotient of the Young modulus of the damaged material by the Young modulus of the undamaged material. Obviously the value of β is between 0 and 1:

(19), $0 \leq \beta \leq 1$,

when $\beta = 1$, the material we consider is undamaged and when $\beta = 0$, the material is completely damaged. The internal quantity β is submitted to the internal constraint (19). We think that this constraint is a material property. Thus it must be taken into account by the elements which define a material. Because it is a constraint on the state, it appears convenient to use the free energy. We do it in the following manner: we decide that the free energy is defined for any value of β, even for the values which are physically impossible. The value of the free energy is $+\infty$ for values of β which are physically impossible, i.e. for $\beta \notin [0,1]$; its value is the usual physical value $\Psi(E)$ when $\beta \in [0,1]$. Thus we have

$$\underline{\Psi}(E) = \Psi(E) + I(\beta),$$

where $\underline{\Psi}$ is the extended free energy defined for any value of β, $\Psi(E)$ is the usual physical value and I is the indicator function of the segment [0,1], ($I(x) = +\infty$ if $x \notin [0,1]$ and $I(x) = 0$ if $x \in [0,1]$), (see the appendix).

Note The temperature T is submitted to the internal constraint $T \geq 0$. For the sake of simplicity, we assume it is always satisfied. If one does not want to make this assumption, it is convenient to add $I_+(T)$ to $\underline{\Psi}$ where I_+ is the indicator function of $[0,+\infty)$.

Before we go on, let us remark that the time derivatives cannot be assumed to be continuous due to the effects of the internal constraint (19). For instance $\frac{d\beta}{dt}$ is discontinuous when β decreases to the value 0 (figure 1).

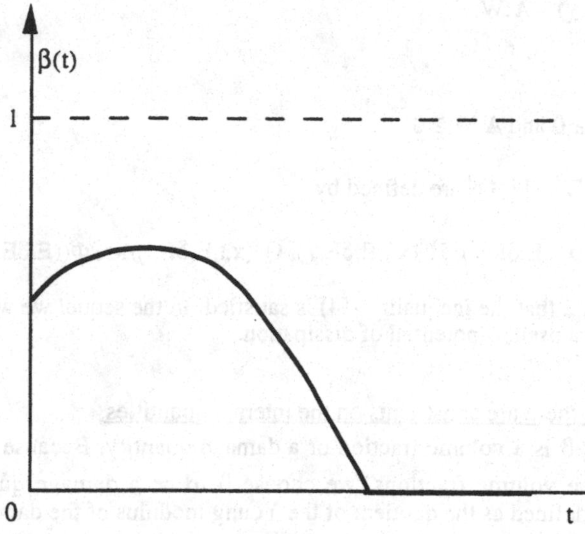

Figure 1

The function β(t) decreases to 0 and remains equal to 0. The derivative $\frac{d\beta}{dt}$ is not a continuous function.

Left and right derivative must be investigated. The right derivative

$$\lim_{\substack{\Delta t \to 0 \\ \Delta t > 0}} \frac{\beta(t+\Delta t) - \beta(t)}{\Delta t} = \frac{d^r\beta}{dt}$$

depends on the future evolution of the material but the left derivative

$$\lim_{\substack{\Delta t \to 0 \\ \Delta t > 0}} \frac{\beta(t) - \beta(t-\Delta t)}{\Delta t} = \frac{d^l\beta}{dt}$$

depends on its past evolution.

The constitutive laws we are looking for must be determinist, i.e. relations between the state quantities E and the history of the material or its past evolution. The left derivative appears as a compulsory choice for any derivative with respect to the time which appears in the constitutive laws. Thus in the sequel <u>all the time derivatives we use are left derivative.</u>

Let us come back at the extended free energy and give one of its properties:

Theorem 3. If the function $\Psi(\varepsilon,\beta,\mathbf{grad}\beta,T)$ is smooth, we have for any actual evolution, i.e. for any evolution such that $\beta(t)\in[0,1]$ at any time t,

$$(20),\quad \forall C\in\partial I(\beta),\quad \frac{d^l\underline{\Psi}}{dt}=\frac{d^l\Psi}{dt}\leq\frac{d^l\Psi}{dt}+C\frac{d^l\beta}{dt},$$

where the derivatives in the formula (20), $\frac{d^l}{dt}$, are left derivatives.

Proof Because the evolution is an actual evolution, we have $I(\beta(t)) = 0$ at any time t. It results the first equality of (20) because $\underline{\Psi}$ is equal to Ψ at any time t. The function $\Psi(\varepsilon,\beta,\mathbf{grad}\beta,T)$ being smooth we have,

$$(21),\quad \Psi(t)-\Psi(t-\Delta t)=\frac{d^l\Psi}{dt}(t)\,\Delta t+o(\Delta t)$$

where the function $o(\Delta t)/\Delta t$ tends to 0 when Δt tends to 0. Because I is a convex function, we have

$$(22),\quad I(\beta(t))-I(\beta(t-\Delta t))\leq C\,(\beta(t)-\beta(t-\Delta t)),\ \text{for any } C\in\partial I(\beta(t)).$$

By dividing (21) and (22) by Δt positive and adding those relations, we get

$$\frac{\Psi(t)-\Psi(t-\Delta t)}{\Delta t}\leq\frac{d^l\Psi}{dt}(t)+C\,\frac{\beta(t)-\beta(t-\Delta t)}{\Delta t}+\frac{o(\Delta t)}{\Delta t}.$$

By letting Δt tend towards 0, we have the inequality of (20).

Because the free energy Ψ is not differentiable whith respect to the time (it has only a left derivative), we decide that all the quantities that are to be derived with respect to the time have left derivatives. We assume also that they are smooth enougth with respect to $\mathbf{x}$ for the calculations to be coherent. The Clausius-Duhem inequality

$$(23),\quad \frac{d\Psi}{dt}+s\frac{dT}{dt}\leq\sigma{:}D(U)+B\frac{d\beta}{dt}+\mathbf{H}.\mathbf{grad}\frac{d\beta}{dt}+\frac{-\mathbf{q}.\mathbf{grad}T}{T},$$

can be proved like in paragraph 7, the time derivatives beeing left derivatives. The choice of left derivative we make is in agreement with the necessity for the the basic inequality (9) or the equivalent Clausius-Duhem inequality to be satisfied by the past evolution of the material. We wish also to keep the role of the Clausius-Duhem inequality as a guide to define the constitutive laws which are relations depending on the past evolution. The left derivative are, as we have already mentionned, deterministic. On the contrary the right derivative are not deterministic.

We use the notations of the preeceding paragraphs. We have the Helmholtz relation

$$(24),\quad s(E)=-\frac{\partial\Psi}{\partial T}(E).$$

Let us define the non dissipative forces. We assume that there exist functions σ^{nd}, B^{nd}, $\mathbf{H}^{nd}$ of E and B^{ndr} of $(E,\mathbf{x},t)$ which satisfy

(13), $\sigma^{nd}(E) = \frac{\partial \Psi}{\partial \varepsilon}(E)$,

(13), $B^{nd}(E) = \frac{\partial \Psi}{\partial \beta}(E)$,

(25), $B^{ndr}(E,x,t) \in \partial I(\beta(x,t))$,

(13), $\mathbf{H}^{nd}(E) = \frac{\partial \Psi}{\partial(\mathbf{grad}\beta)}(E)$.

With those definitions we can write the Clausius-Duhem inequality as in the previous paragraph

$$(\sigma-\sigma^{nd}):D(U) + (B-B^{nd})\frac{d\beta}{dt} + (\mathbf{H}-\mathbf{H}^{nd}).\mathbf{grad}\frac{d\beta}{dt} + T\mathbf{q}.\mathbf{grad}\frac{1}{T} \geq 0.$$

The relations (13), (24) and (25) are the state laws. It results from the preeceding formulae that the smooth part of the extented free energy is differentiable and that the non-smooth part is subdifferentiable, i.e. that the subdifferential set $\partial I(\beta)$ is not empty. Let us see how important is this assumption. The quantity $B^{ndr}(E,x,t)$ is the thermodynamical reaction to the internal constraint (19). It is related to β by the state law (25). This one implies that the subdifferential $\partial I(\beta)$ is not empty, thus that β is between 0 and 1 which means that the internal constraint (19) is satisfied. One can also say that relation (25) has two meanings, first that the internal constraint (19) is satisfied, second that there exists a reaction to the internal constraint which is zero for $0 < \beta < 1$, positive for $\beta = 1$ and negative for $\beta = 0$. Let us also note that the sum of the reaction B^{ndr} and of the reversible force B^{nd}, $(B^{ndr}+ B^{nd})$ is a generalized derivative of the free energy $\underline{\Psi}$ with respect to β; B^{nd} is the smooth part and B^{ndr} is the non-smooth part of the derivative. If the indicator function I is approximated by a smooth function, B^{ndr} is approximated by a classical derivative and there is no more difference between the smooth part B^{nd} and the non-smooth part B^{ndr}. In our point of view, the non-smooth mechanics point of view, the free energy is $\underline{\Psi}$ and the non-dissipative force associated to β is $B^{ndr}+ B^{nd} \in \partial\underline{\Psi}$.

<u>Note</u> If we do not assume that the temperature is positive, we replace the relation (24) by :

the entropy is a function of (E,x,t) which satisfies

$$s(E,x,t) \in -\frac{\partial \Psi}{\partial T}(E) + \partial I_+(T).$$

This relation implies that $\partial I_+(T)$ is not empty. It results that the temperature T is positive. It shows also that if the temperature is strictly positive the classical Helmholtz relation (24) is verified.

To complete the description of the constitutive laws, we assume the assumption (14) made in paragraph 5 is satisfied : there exist four functions of E and of $f \in S$, $c \in \mathbf{R}$, $\mathbf{g} \in \mathbf{R}^3$, $\mathbf{p} \in \mathbf{R}^3$ and other quantities χ depending on the history of the material: σ^d, $\mathbf{B}^d$, $\mathbf{H}^d$, and $\mathbf{Q}^d$ which satisfy

$$\forall \mathbf{x},\mathbf{t}, \forall \mathrm{E}, \forall \mathrm{W} = (\mathrm{f,c,}\mathbf{g},\mathbf{p}) \in \mathrm{SxRxR^3xR^3}, \forall \chi,$$

(14), $\sigma^d(\mathbf{x},\mathbf{t},\mathrm{E},\mathrm{W},\chi):\mathrm{f} + \mathrm{B}^d(\mathbf{x},\mathbf{t},\mathrm{E},\mathrm{W},\chi)\mathrm{c} + \mathbf{H}^d(\mathbf{x},\mathbf{t},\mathrm{E},\mathrm{W},\chi).\mathbf{g} + \mathbf{Q}^d(\mathbf{x},\mathbf{t},\mathrm{E},\mathrm{W},\chi).\mathbf{p} \geq 0.$

Then the constitutive laws are defined by the following relations

(15), $\sigma = \sigma^d(\mathbf{x},\mathbf{t},\mathrm{E},\delta\mathrm{E},\chi) + \sigma^{nd}(\mathrm{E}),$

(26), $\mathrm{B} = \mathrm{B}^d(\mathbf{x},\mathbf{t},\mathrm{E},\delta\mathrm{E},\chi) + \mathrm{B}^{nd}(\mathrm{E}) + \mathrm{B}^{ndr}(\mathrm{E},\mathbf{x},\mathbf{t}),$

(17), $\mathbf{H} = \mathbf{H}^d(\mathbf{x},\mathbf{t},\mathrm{E},\delta\mathrm{E},\chi) + \mathbf{H}^{nd}(\mathrm{E}),$

(18), $\mathrm{T}\mathbf{q} = \mathbf{Q}^d(\mathbf{x},\mathbf{t},\mathrm{E},\delta\mathrm{E},\chi),$

where $\mathrm{E} = (\varepsilon,\beta,\mathbf{grad}\beta,\mathrm{T})$ et $\delta\mathrm{E} = (\frac{d^l\varepsilon}{dt}, \frac{d^l\beta}{dt}, \mathbf{grad}\frac{d^l\beta}{dt}, \mathbf{grad}\frac{1}{T})$.

The functions σ^d, B^d, $\mathbf{H}^d$ and $\mathbf{Q}^d$ are the dissipative or irreversible forces. Let us note that the constitutive laws are obviously deterministic because the time derivatives are left derivatives. We must prove that our choice is such that the Clausius-Duhem inequality is satisfied. The following theorem shows that we have the expected properties.

<u>Theorem 4.</u> If the state laws (13), (24) and (25), the constitutive laws (15), (26), (17) and (18) and the inequality (14) are satisfied then,

(i) the internal constraint (19) is satisfied;

(ii) the Clausius-Duhem rellation (23) is satisfied.

<u>Proof</u> We have already seen that the state law (25) implies that the the internal constraint (19) is satisfied because the subdifferential set $\partial I(\beta)$ is not empty. Let us replace W by $\delta\mathrm{E} = (\frac{d^l\varepsilon}{dt}, \frac{d^l\beta}{dt}, \mathbf{grad}\frac{d^l\beta}{dt}, \mathbf{grad}\frac{1}{T})$ in the inequality (14). Let us replace also the dissipative forces by their expression given by the constitutive laws (15), (26), (17) and (18). We get

$$(\sigma - \sigma^{nd}(\mathrm{E})) : \frac{d^l\varepsilon}{dt} + (\mathrm{B} - \mathrm{B}^{nd}(\mathrm{E}) - \mathrm{B}^{ndr}(\mathrm{E},\mathbf{x},\mathbf{t}))\frac{d^l\beta}{dt} + (\mathbf{H} - \mathbf{H}^{nd}(\mathrm{E})).\mathbf{grad}\frac{d^l\beta}{dt}$$

$$+\mathrm{T}\mathbf{q}.\mathbf{grad}\frac{1}{T} \geq 0.$$

This relation gives with the state laws (13), (24) and (25),

(27), $\sigma:\frac{d^l\varepsilon}{dt} + \mathrm{B}\frac{d^l\beta}{dt} + \mathbf{H}.\mathbf{grad}\frac{d^l\beta}{dt} - \frac{d^l\Psi}{dt} - \mathrm{B}^{ndr}(\mathrm{E},\mathbf{x},\mathbf{t})\frac{d^l\beta}{dt} + \mathrm{T}\mathbf{q}.\mathbf{grad}\frac{1}{T} - s\frac{d^lT}{dt} \geq 0.$

Because the assumptions of theorem 3 are satisfied, the relation (20) with $\mathrm{C} = \mathrm{B}^{ndr}(\mathrm{E},\mathbf{x},\mathbf{t})$ gives

$$-\frac{d^l\Psi}{dt} - B^{ndr}(E,x,t)\frac{d^l\beta}{dt} \leq -\frac{d^l\Psi}{dt}.$$

This inequality and (27) give

$$\sigma:\frac{d^l\varepsilon}{dt} + B\frac{d^l\beta}{dt} + \mathbf{H}.\mathbf{grad}\frac{d^l\beta}{dt} + T\mathbf{q}.\mathbf{grad}\frac{1}{T} - s\frac{d^lT}{dt} \geq \frac{d^l\Psi}{dt},$$

which is the Clausius-Duhem inequality (23) with left derivatives.

Let us sum up. In this theory, a material is defined by choosing

the state quantities E,

the linear spaces $\mathbb{V}$ and $\mathbb{C}$,

the power of the internal forces $P_i(D,\mathbb{V},c)$,

the quantities χ,

the function $\underline{\Psi}$,

the functions σ^d, B^d, $\mathbf{H}^d$ and $\mathbf{Q}^d$ or the pseudo-potential of dissipation Φ.

8.1. <u>Constitutive laws on discontinuity surfaces.</u>

In this paragraph we do not make the small perturbation assumption. The state quantities or the velocities can be discontinuous on some surface. It is easy to prove that the equations of movement on those surfaces are,

$$m[\mathbf{U}] = [\,\sigma.\mathbf{N}],$$
$$[\mathbf{H}.\mathbf{N}] = 0,$$

where m is the mass flux, $\mathbf{N}$ is a unit normal vector to the discontinuity line which divides the spaces into two parts denoted with the indices 1 and 2; $[Z] = Z_2 - Z_1$ denotes the discontinuity of the quantity Z. The energy balance is

$$m[e'] - \underline{\mathbf{T}}.[\mathbf{U}] - h[b] = [-\mathbf{q}.\mathbf{N}],$$

where e' is the specific internal energy ($e = \rho e'$, ρ is the density)

$$h = \mathbf{H}.\mathbf{N},\ \underline{\mathbf{T}} = \frac{1}{2}(\mathbf{T}_1 + \mathbf{T}_2),\ Q = -\mathbf{q}.\mathbf{N}.$$

Let us recall that b is the actual velocity of β, $b = \frac{d\beta}{dt}$. By using the mass balance, we get,

$$m([e'] - \underline{T}_N\,[\frac{1}{\rho}]) - \underline{\mathbf{T}}_T.[\mathbf{U}_T] - h[b] - [Q] = 0,$$

where

$$\underline{T}_N = \underline{\mathbf{T}}.\mathbf{N},\ \underline{\mathbf{T}}_T = \underline{\mathbf{T}} - \underline{T}_N\mathbf{N} \text{ and } \mathbf{U}_T = \mathbf{U} - U_N\mathbf{N} \text{ is the tangential velocity.}$$

The second principle of thermodynamics is

(28), $m[s'] - [\frac{Q}{T}] \geq 0,$

where s' is the specific entropy ($s = \rho s'$) or

$$mT_h[s'] - T_h \underline{Q}[\frac{1}{T}] - [Q] \geq 0,$$

where T_h is the harmonic average temperature ($\frac{1}{T_h} = \frac{1}{2}(\frac{1}{T_1} + \frac{1}{T_2})$) and Q_m the average heat flux ($Q_m = \frac{1}{2}(Q_1+Q_2)$). It gives with the energy balance

$$([T_h s' - e'] + \underline{T}_N[\frac{1}{\rho}])m - T_h Q_m[\frac{1}{T}] + \underline{T}_T.[U_T] + h[b] \geq 0.$$

To describe the evolution of the discontinuity surface constitutive laws are needed. They are defined by assuming that there exist four functions of $\Delta E = (m,[\frac{1}{T}],[U_T],[b])$ and of (x,t), $E = (E_1,E_2)$ and other quantities ξ depending on the history of the material : Y^d, Q^d, F^d, and h^d which satisfy

$$\forall x,t,\ \forall E,\ \forall W = (f,g,V,c) \in \mathbf{R x R x R^2 x R}, \forall \xi,$$

(29), $Y^d(x,t,E,W,\xi)f + Q^d(x,t,E,W,\xi)g + F^d(x,t,E,W,\xi).V + h^d(x,t,E,W,\xi)c \geq 0.$

The constitutive laws are

$$[T_h s' - e'] + \underline{T}_N[\frac{1}{\rho}] = Y^d(x,t,E,\Delta E,\xi),$$
$$-T_h Q_m = Q^d(x,t,E,\Delta E,\xi),$$
$$\underline{T}_T = F^d(x,t,E,\Delta E,\xi),$$
$$h = h^d(x,t,E,\Delta E,\xi).$$

It is easy to prove that the preceeding constitutive laws are such that the second principle is satisfied:

<u>Theorem 5</u> If the inequality (29) is satisfied, the fondamental inequality (28) is satisfied in any actual evolution such that the temperature is strictly positive.

<u>Proof.</u> Write the inequality (29) with the actual velocities $W = \Delta E = (m,[\frac{1}{T}],[U_T],[b])$ and substract the energy balance to get the fondamental inequality multiplied by the harmonic average temperature.

Let us sum up again. In this theory, a material is defined by choosing

the state quantities E,

the linear spaces $\mathbb{V}$ and $\mathbb{C}$,

the power of the internal forces $P_i(D,\mathbf{V},c)$,

the quantities χ,

the function Ψ,

the functions σ^d, $\mathbf{B}^d$, $\mathbf{H}^d$ and $\mathbf{Q}^d$ or the pseudo-potential of dissipation Φ,

the quantities ξ,

the functions Y^d, Q^d, $\mathbf{F}^d$, and h^d or the pseudo-potential of dissipation on the discontinuity surface

8.1.1. Example of constitutive laws on discontinuity lines.

We choose pseudo-potential of dissipation $\Phi_s(\Delta E) = I_{0r}([\frac{1}{T}])$ where I_{0r} is the indicator function of $\mathbf{R}\mathrm{x}\{0\}\mathrm{x}\mathbf{R}^2\mathrm{x}\mathbf{R}$. It gives the constitutive laws,

$$Y^d(\Delta E) = 0,\ Q^d(\mathbf{x},t,E,\Delta E) \in \partial I_{0r}([\frac{1}{T}]),\ \mathbf{F}^d(\Delta E) = 0 \text{ and } h^d(\Delta E) = 0.$$

It results from the second constitutive law that the temperature is continuous. If we assume that the density is continuous, the first constitutive law gives

$$[Ts' - e'] = -[\Psi'] = 0.$$

The two last laws show that there is no friction and no flux of damage work on the discontinuity surface. These constitutive laws are often chosen to describe phase changes occuring in solids.

9. Shape memory alloys. A macroscopic theory.

9.1. Introduction.

This chapter is devoted to the construction of models able to describe at the macroscopic level the evolution of a structure made of shape memory alloys. The internal constraints on the state quantites will play a major role and account for most of the striking properties of the shape memory alloys.

9.2. The state quantities.

As already mentionned we deal only with macroscopic phenomenons and macroscopic quantities. Thus to describe the deformations of the alloy, we choose the macroscopic deformation ε. For the sake of simplicity we assume this deformation to be small (a large deformation theory based on those ideas exists). Of course the temperature T is a thermodynamical quantity.

The properties of shape memory alloys results from martensite-austenite phase changes produced either by thermal actions (as it is usual) or by mechanical actions. At the macroscopic level we need quantities to describe those phase changes. For this purpose we choose as new thermodynamical quantities the volume fractions β_i of the martensite and austenite. We think that this choice is the more simple we can make. Again to be very simple we assume that only two martensites exist together with the austenite. The volume fractions of the martensites are β_1 and β_2. The volume fraction of austenite is β_3. Those volume fractions are not independant: they satisfy constraints, said as usual internal constraints,

(30), $0 \le \beta_i \le 1$,

because the β's are volumetric proportions. We assume that no void can appear in the evolutions of the alloy, i. e. $\beta_1+\beta_2+\beta_3 \ge 1$ and that no interpenetration of the phases can occur, i. e. $\beta_1+\beta_2+\beta_3 \le 1$. Thus the β's satisfy an other internal constraint,

(31), $\beta_1+\beta_2+\beta_3 = 1$.

We think those internal constraints are physical properties.

The thermodynamical macroscopic state quantities we have chosen are $E = (\varepsilon,\beta_1,\beta_2,\beta_3,T)$ or $E = (\varepsilon,\beta_1,\beta_2,\beta_3,\mathbf{grad}\beta_1,\mathbf{grad}\beta_2,\mathbf{grad}\beta_3,T)$ depending on the sophistication we wish. The second set $(\varepsilon,\beta_1,\beta_2,\beta_3,\mathbf{grad}\beta_1,\mathbf{grad}\beta_2,\mathbf{grad}\beta_3,T)$ is chosen if we think that the composition of the alloy at one point is influenced by its neighbourhood. We note β the vector (β_i).

9.3. <u>The free energy.</u>

As already said, we consider a shape memory alloy as a mixture of the three martensite austenite phases with volume fractions β_i. The volumetric energy of the mixture we choose is

(32), $$\Psi(E) = \sum_{i=1}^{3} \beta_i \Psi_i(E) + Th(\beta)$$

where the Ψ_i are the volumetric free energies of the i phases and Th is a free energy describing interactions between the different phases. We have said that internal constraints are physical properties. Being physical properties we decide to take them into account with the two functions we have to describe the material, i.e. the free energy Ψ and the pseudo-potential of dissipation Φ. The pseudo-potential describes the kinematic properties, i.e, properties which depend on the velocities. The free energy describes the state properties. Obviously the internal constraints (30) and (31) are not kinematic properties. Thus we take them into account with the free energy Ψ. For this purpose, we assume the Ψ_i to be defined over the whole linear space spanned by the ε, β_i and $\mathbf{grad}\beta_i$ and define the extended free energy by

$$\underline{\Psi}(E) = \Psi(E) + TI_0(\beta) = \Psi(E) + I_0(\beta) = \sum_{i=1}^{3} \beta_i \Psi_i(E) + T\underline{h}(\beta),$$

where I_0 is the indicator function of the convex set

$$C = \{(\gamma_1,\gamma_2,\gamma_3) \in \mathbf{R}^3;\ 0 \le \gamma_i \le 1\ ;\ \gamma_1+\gamma_2+\gamma_3 = 1\},$$

and the extended interaction free energy is defined by

$\underline{h}(\beta) = h(\beta)+I_0(\beta).$

The more simple choice for $h(\beta)$ we can make is $h(\beta) = 0$. There is no interaction betwween the different phases in the mixture. The extended interaction free energy $\underline{h}(\beta) = I_0(\beta)$ is equal to 0 when the mixture is physically possible ($\beta \in C$) and to $+\infty$ when the mixture is physically impossible ($\beta \notin C$). Properties of the extended free energies are given in the theorem,

Theorem 6. If the function $\Psi(\varepsilon,\beta,\mathbf{grad}\beta,T)$ is smooth, we have for any actual evolution, i.e. for any evolution such that $\beta \in C$ at any time t,

$$(33),\quad \forall \mathbf{B} \in \partial I_0(\beta),\quad \frac{d^l\underline{\Psi}}{dt} = \frac{d^l\Psi}{dt} \leq \frac{d^l\Psi}{dt} + \mathbf{B}.\frac{d^l\beta}{dt} = \frac{d^l\Psi}{dt} + B_1\frac{d^l\beta_1}{dt} + B_2\frac{d^l\beta_2}{dt} + B_3\frac{d^l\beta_3}{dt},$$

where the derivatives in the formula (33), $\frac{d^l}{dt}$, are left derivatives.

Proof. It is identical to the proof of theorem 3.

The vector $\mathbf{B}$ is the thermodynamical reaction to the internal constraints (30) and (31). The subdifferential of the indicator function I_0 is rather easily computed :

Theorem 7. The subdifferential ∂I_0 is

$\partial I_0(\beta) = (c,c,c)$, if β is an internal point of C ($0 < \beta_i < 1$ for any i);

$\partial I_0(0,\beta_2,\beta_3) = (-a^2 + c,c,c)$, if $0 < \beta_i < 1$ for i = 2,3;

$\partial I_0(0,0,1) = (-a^2 + c,-b^2 + c,c)$;

where a, b and c are real numbers.

Proof. Let I_i be the indicator function of the set $\{(\gamma_1,\gamma_2,\gamma_3) \in \mathbf{R}^3 ;\ 0 \leq \gamma_i \leq 1\}$ and I_4 the indicator function of the set $\{(\gamma_1,\gamma_2,\gamma_3) \in \mathbf{R}^3;\ \gamma_1+\gamma_2+\gamma_3 = 1\}$. We have $I_0(\beta) = I_1(\beta)+I_2(\beta)+I_3(\beta)+I_4(\beta)$. It results from a theorem of convex analysis (see for instance Moreau [20]) that

$\partial I_0(\beta) = \partial I_1(\beta)+\partial I_2(\beta)+\partial I_3(\beta)+\partial I_4(\beta),$

(be careful, this result obvious for smooth functions is not always true for convex non-smooth functions).

When $0 < \beta_i < 1$ for any i, we have $\partial I_0(\beta) = \partial I_4(\beta) = (c,c,c)$.

When $\beta_1 = 0$, $0 < \beta_i < 1$ for i = 2, 3, we have $\partial I_0(\beta) = \partial I_1(\beta)+\partial I_4(\beta) = (-a^2,0,0)+(c,c,c)$.

When $\beta_1 = \beta_2 = 0$, we have $\partial I_0(\beta) = \partial I_1(\beta)+\partial I_2(\beta)+\partial I_4(\beta) = (-a^2,0,0)+(0,-b^2,0)+(c,c,c)$.

Note. It is also possible to prove the theorem by using the fact that the thermodynamical reaction $\mathbf{B}$ is a vector of $\mathbf{R}^3$ normal to the convex set C of $\mathbf{R}^3$.

As for the volumic free energies we choose,

$$\Psi_1(E) = \frac{1}{2}\varepsilon^T{:}K_1\varepsilon + \sigma_1(T)^T{:}\varepsilon - C_1 T \mathrm{Log} T,$$
$$\Psi_2(E) = \frac{1}{2}\varepsilon^T{:}K_2\varepsilon + \sigma_2{}^T(T){:}\varepsilon - C_2 T \mathrm{Log} T,$$
$$\Psi_3(E) = \frac{1}{2}\varepsilon^T{:}K_3\varepsilon - \frac{l_a}{T_0}(T-T_0) - C_3 T \mathrm{Log} T,$$

where K_i are the elastic tensors, C_i the heat capacities of the phases. The stresses σ_1 and σ_2 depend on the temperature T. The quantity l_a is roughthy the martensite-austenite phase change latent heat (see paragraph 9.8.3). We denote $\sigma^T{:}\varepsilon = \sigma_{ij}\varepsilon_{ij}$.

For the interaction function we choose

$$T\underline{h}(E) = I_0(\beta) \text{ or } T\underline{h}(E) = I_0(\beta) + \frac{k}{2}(\mathbf{grad}\beta_1)^2 + \frac{k}{2}(\mathbf{grad}\beta_2)^2,$$

depending on the sophistication of the model.

Because we want to describe the main basic features of the shape memory alloys behaviour, we assume for the sake of simplicity that the elastic tensors K_i and the heat capacities C_i are the same for all the phases:

$$C_i = C, \qquad K_i = K, \qquad \text{for } i = 1, 2 \text{ and } 3.$$

Always for the sake of simplicity we assume that

$$\sigma_1(T) = -\sigma_2(T) = -\tau(T).$$

Concerning the stress $\tau(T)$, we know that at high temperature the behaviour of the alloy is a classical elastic behaviour. Thus we have $\tau(T) = 0$ at high temperature and choose the schematic simple expression (always for the sake of simplicity),

$$\tau(T) = (T-T_c)\underline{\tau}, \text{ for } T \le T_c,$$
$$\tau(T) = 0, \text{ for } T \ge T_c,$$

with $\underline{\tau}_{11} \le 0$ and assume the temperature T_c to be greater than T_0. With those assumption we get

$$\Psi(E) = \frac{1}{2}\varepsilon^T{:}K\varepsilon - (\beta_1-\beta_2)\tau(T)^T{:}\varepsilon - \beta_3\frac{l_a}{T_0}(T-T_0) - CT\mathrm{Log}T.$$

We have the Helmholtz relation

$$s(E) = -\frac{\partial\Psi}{\partial T}(E) = (\beta_1-\beta_2)\underline{\tau}^T{:}\varepsilon + \beta_3\frac{l_a}{T_0} + C(1+\mathrm{Log}T), \text{ for } T \le T_c,$$

(24). $$s(E) = -\frac{\partial\Psi}{\partial T}(E) = \beta_3\frac{l_a}{T_0} + C(1+\mathrm{Log}T), \text{ for } T \ge T_c,$$

which gives the volumic internal energy,

$$e(E) = \Psi(E)+Ts(T) = \frac{1}{2}\varepsilon^T{:}K\varepsilon + (\beta_1-\beta_2)T_c\underline{\tau}^T{:}\varepsilon + \beta_3 l_a + CT, \text{ for } T \leq T_c,$$

$$e(E) = \Psi(E)+Ts(T) = \frac{1}{2}\varepsilon^T{:}K\varepsilon + \beta_3 l_a + CT, \text{ for } T \geq T_c.$$

Note When $T = T_c$, the energy can be discontinuous. On the discontinuity surface the equations of paragraph 8.1 apply. If one wants to avoid using them, the free energy function Ψ can be smoothed for the value $T = T_c$ to have the entropy continuous.

The non dissipative forces are defined by assuming that there exist functions σ^{nd}, $\mathbf{B}^{nd}$, $\mathbf{H}^{nd}$ of E and $\mathbf{B}^{ndr}$ of (E,x,t) which satisfy

$$(13), \quad \sigma^{nd}(E) = \frac{\partial\Psi}{\partial\varepsilon}(E) = K\varepsilon-(\beta_1-\beta_2)\tau(T),$$

$$(13), \quad \mathbf{B}^{nd}(E) = \frac{\partial\Psi}{\partial\beta}(E) = \begin{vmatrix} -\tau(T)^T{:}\varepsilon \\ \tau(T)^T{:}\varepsilon \\ \frac{l_a}{T_0}(T-T_0) \end{vmatrix},$$

$$(25), \quad \mathbf{B}^{ndr}(E,\mathbf{x},t) \in \partial I(\beta(\mathbf{x},t)),$$

$$(13), \quad \mathbf{H}^{nd}(E) = \frac{\partial\Psi}{\partial(\mathbf{grad}\beta)}(E) = \begin{vmatrix} k\mathbf{grad}\beta_1 \\ k\mathbf{grad}\beta_2 \\ 0 \end{vmatrix}.$$

To complete the description of the constitutive laws, we assume the assumption (14) made in paragraph 7 is satisfied: there exist four functions of E and of x,t, $f \in S$, $c \in \mathbf{R}^3$, $g \in \mathbf{R}^{3x3}$ and $p \in \mathbf{R}^3$ and other quantities χ depending on the history of the material : σ^d, $\mathbf{B}^d$, $\mathbf{H}^d$, and $\mathbf{Q}^d$ which satisfy,

$$\forall \mathbf{x},t\ \forall E,\ \forall W = (f,c,g,p) \in S x \mathbf{R}^3 x \mathbf{R}^{3x3} x \mathbf{R}^3,\ \forall\chi,$$

$$(14), \quad \sigma^d(\mathbf{x},t,E,W,\chi){:}f + \mathbf{B}^d(\mathbf{x},t,E,W,\chi)c + \mathbf{H}^d(\mathbf{x},t,E,W,\chi).g + \mathbf{Q}^d(\mathbf{x},t,E,W,\chi).\mathbf{p} \geq 0.$$

Then the constitutive laws are defined by the following relations,

$$(15), \quad \sigma = \sigma^d(\mathbf{x},t,E,\delta E,\chi) + \sigma^{nd}(E),$$

$$(26), \quad \mathbf{B} = \mathbf{B}^d(\mathbf{x},t,E,\delta E,\chi) + \mathbf{B}^{nd}(E) + \mathbf{B}^{ndr}(E,\mathbf{x},t),$$

$$(17), \quad \mathbf{H} = \mathbf{H}^d(\mathbf{x},t,E,\delta E,\chi) + \mathbf{H}^{nd}(E),$$

$$(18), \quad T\mathbf{q} = \mathbf{Q}^d(\mathbf{x},t,E,\delta E,\chi),$$

where $E = (\varepsilon, \beta, \mathbf{grad}\beta, T)$ and $\delta E = (\frac{d^i\varepsilon}{dt}, \frac{d^i\beta}{dt}, \mathbf{grad}\frac{d^i\beta}{dt}, \mathbf{grad}\frac{1}{T})$.

9.4. The non-dissipative constitutive laws.

We assume that there is no mechanical dissipation (the functions σ^d, $\mathbf{B}^d$ and $\mathbf{H}^d$ are equal to 0) or that the non-thermal part of the pseudo-potential of dissipation is equal to 0. We know that this is not very realistic but it is a step towards the complete understanting of the constitutive laws. The results from this non-dissipative theory will be schematic. The constitutive laws are given by

(34), $\sigma = \sigma^{nd}(E) = \frac{\partial\Psi}{\partial\varepsilon}(E)$,

(35), $\mathbf{B} = \mathbf{B}^{nd}(E) = \frac{\partial\Psi}{\partial\beta}(E) + \mathbf{B}^{ndr}(E,\mathbf{x},t)$,

(36), $\mathbf{B}^{ndr}(E,\mathbf{x},t) \in \partial I_0(\beta(\mathbf{x},t))$,

(37), $\mathbf{H} = \mathbf{H}^{nd}(E) = \frac{\partial\Psi}{\partial(\mathbf{grad}\beta)}(E)$,

(38), $T\mathbf{q} = \mathbf{Q}^d(E,\delta E,\chi)$,

The last constitutive law gives the classical Fourier's law

$$\mathbf{q} = -\lambda\mathbf{grad}T,$$

where λ is the thermal conductivity, by choosing $\mathbf{Q}^d(E,\delta E,\chi) = \lambda T^3\,\mathbf{grad}\frac{1}{T}$.

9.5. Transformation of the equations. Elimination of β_3.

Because of the relation (31), we can select one of the β's, for instance $\beta_3 = (1-\beta_1-\beta_2)$ and rewrite all the equations with the dissymetric set of state quantities $E_r = (\varepsilon,\beta_1,\beta_2,T)$ or $E_r = (\varepsilon,\beta_1,\beta_2,\mathbf{grad}\beta_1,\mathbf{grad}\beta_2,T)$.

Let us define

$$\beta_r = (\beta_1,\beta_2),$$

and the convex set C_r of $\mathbf{R}^2$ by

$$C_r = \{(\gamma_1,\gamma_2,)\in \mathbf{R}^2;\ 0 \le \gamma_i \le 1\ ;\ \gamma_1+\gamma_2 \le 1\},$$

and note that $\gamma\in C$ is equivalent to $\gamma_r = (\gamma_1,\gamma_2)\in C_r$ and $\gamma_3 = 1-\gamma_1-\gamma_2$ or to $I_0(\gamma) = I_r(\gamma_r) = 0$, and $\gamma_3 = 1-\gamma_1-\gamma_2$ where I_r is the indicator function of C_r. Let us remark that $\mathbf{B}^{ndr}\in \partial I_0(\beta)$ is equivalent to

(39), $\beta\in C$ and $\forall\gamma\in C,\ 0 \ge \mathbf{B}^{ndr}(\gamma-\beta) = B_1^{ndr}(\gamma_1-\beta_1) + B_2^{ndr}(\gamma_2-\beta_2) + B_3^{ndr}(\gamma_3-\beta_3)$.

Thus the relation (39) is equivalent to

$$\beta_r = (\beta_1,\beta_2)\in C_r,\ \forall\gamma\in C_r,\ 0 \geq (B_1-B_3)\gamma_1+(B_2-B_3)\gamma_2,$$

or to

$$\mathbf{B}_r^{ndr}\in \in \partial I_r(\beta_r) \text{ with } \mathbf{B}_r^{ndr} = \begin{vmatrix} B_1^{ndr}-B_3^{ndr} \\ B_2^{ndr}-B_3^{ndr} \end{vmatrix},$$

where I_r is the indicator function of the convex set C_r. Thus the equation (36) is equivalent to

(40), $\mathbf{B}_r^{ndr}\in \partial I_r(\beta_r)$.

The power of the internal forces

$$P_i(D,\mathbf{U},\frac{d\beta_1}{dt},\frac{d\beta_2}{dt},\frac{d\beta_3}{dt},\mathbf{grad}\frac{d\beta_1}{dt},\mathbf{grad}\frac{d\beta_2}{dt},\mathbf{grad}\frac{d\beta_3}{dt}) = -\int_D \sigma{:}D(\mathbf{U})d\Omega$$

$$-\int_D\{B_1\frac{d\beta_1}{dt}+B_2\frac{d\beta_2}{dt}+B_3\frac{d\beta_3}{dt}+\mathbf{H}_1.\mathbf{grad}\frac{d\beta_1}{dt}+\mathbf{H}_2.\mathbf{grad}\frac{d\beta_2}{dt}+\mathbf{H}_3.\mathbf{grad}\frac{d\beta_3}{dt}\}d\Omega,$$

becomes

$$P_i(D,\mathbf{U},\frac{d\beta_1}{dt},\frac{d\beta_2}{dt},\mathbf{grad}\frac{d\beta_1}{dt},\mathbf{grad}\frac{d\beta_2}{dt}) = -\int_D \sigma{:}D(\mathbf{U})d\Omega$$

$$-\int_D\{(B_1-B_3)\frac{d\beta_1}{dt}+(B_2-B_3)\frac{d\beta_2}{dt}+(\mathbf{H}_1-\mathbf{H}_3).\mathbf{grad}\frac{d\beta_1}{dt}+(\mathbf{H}_2-\mathbf{H}_3).\mathbf{grad}\frac{d\beta_2}{dt}\}d\Omega$$

$$= -\int_D \sigma{:}D(\mathbf{U})d\Omega - \int_D\{\mathbf{B}_r.\frac{d\beta_r}{dt}+\mathbf{H}_r.\mathbf{grad}\frac{d\beta_r}{dt}\}d\Omega,$$

with

$$\mathbf{B}_r = \begin{vmatrix} B_1-B_3 \\ B_2-B_3 \end{vmatrix} \text{ and } \mathbf{H}_r = \begin{vmatrix} \mathbf{H}_1-\mathbf{H}_3 \\ \mathbf{H}_2-\mathbf{H}_3 \end{vmatrix}.$$

The power of the external forces

$$P_e(D,\mathbf{U},\frac{d\beta_1}{dt},\frac{d\beta_2}{dt},\frac{d\beta_3}{dt}) = \int_D \mathbf{f}.\mathbf{U}d\Omega$$

$$+\int_D\{A_1\frac{d\beta_1}{dt}+A_2\frac{d\beta_2}{dt}+A_3\frac{d\beta_3}{dt}\}d\Omega+\int_{\partial D}\mathbf{T}.\mathbf{U}d\Gamma+\int_{\partial D}\{a_1\frac{d\beta_1}{dt}+a_2\frac{d\beta_2}{dt}+a_3\frac{d\beta_3}{dt}\}d\Gamma,$$

becomes

$$P_e(D,\mathbf{U},\frac{d\beta_1}{dt},\frac{d\beta_2}{dt}) = \int_D \mathbf{f}.\mathbf{U}d\Omega$$

$$+ \int_D \{(A_1-A_3)\frac{d\beta_1}{dt} + (A_2-A_3)\frac{d\beta_2}{dt}\}d\Omega + \int_{\partial D} \mathbf{T}.\mathbf{U}d\Gamma + \int_{\partial D} \{(a_1-a_3)\frac{d\beta_1}{dt} + (a_2-a_3)\frac{d\beta_2}{dt}\}d\Gamma.$$

$$= \int_D \mathbf{f}.\mathbf{U}d\Omega + \int_D \mathbf{A}_r.\frac{d\beta_r}{dt}d\Omega + \int_{\partial D} \mathbf{T}.\mathbf{U}d\Gamma + \int_{\partial D} \mathbf{a}_r.\frac{d\beta_r}{dt}\,d\Gamma,$$

with

$$\mathbf{A}_r = \begin{vmatrix} A_1-A_3 \\ A_2-A_3 \end{vmatrix}, \mathbf{a}_r = \begin{vmatrix} a_1-a_3 \\ a_2-a_3 \end{vmatrix}.$$

We define also

$$h_r(E_r) = h(E),\ \Psi_{ir}(E_r) = \Psi_i(E), \text{ with } \beta_3 = (1-\beta_1-\beta_2),$$

$$\Psi_r(E_r) = \beta_1\Psi_{1r}(E_r)+\beta_2\Psi_{2r}(E_r)+(1-\beta_1-\beta_2)\Psi_{3r}(E_r)+Th_r(E_r),$$

$$\underline{\Psi}_r(E_r) = \Psi_r(E_r)+I_r(\beta_r) = \beta_1\Psi_{1r}(E_r)+\beta_2\Psi_{2r}(E_r)+(1-\beta_1-\beta_2)\Psi_{3r}(E_r)+T\underline{h}_r(E_r),$$

with

$$T\underline{h}_r(E_r) = Th_r(E_r)+I_r(\beta_r).$$

The expressions of the different powers show that equations of the movement,

(5), $0 = \text{div}\mathbf{H} - \mathbf{B} + \mathbf{A}$, in D,

(6), $\mathbf{H}.\mathbf{N} = \mathbf{a}$, in ∂D,

become,

(5ter), $0 = \text{div}\mathbf{H}_r - \mathbf{B}_r + \mathbf{A}_r$, in D,

(6ter), $\mathbf{H}_r.\mathbf{N} = \mathbf{a}_r$, in ∂D.

In the case where the power of the internal forces does depend on the gradients the equation (5bis) becomes

(5quatro), $0 = -\mathbf{B}_r + \mathbf{A}_r$, in D.

The non dissipative internal forces are

$$\sigma_r^{nd}(E_r) = \frac{\partial\Psi_r}{\partial\varepsilon}(E_r) = \sigma^{nd}(E) = \frac{\partial\Psi}{\partial\varepsilon}(E), \text{ with } \beta_3 = (1-\beta_1-\beta_2),$$

$$\mathbf{B}_r{}^{nd}(E_r) = \frac{\partial \Psi_r}{\partial \beta_r}(E_r) = \mathbf{Y}_r(E_r) = \begin{vmatrix} \Psi_{1r}(E_r)-\Psi_{3r}(E_r) \\ \Psi_{2r}(E_r)-\Psi_{3r}(E_r) \end{vmatrix} = \begin{vmatrix} -\tau(T)^T:\varepsilon-\frac{l_a}{T_0}(T-T_0) \\ \tau(T)^T:\varepsilon-\frac{l_a}{T_0}(T-T_0) \end{vmatrix},$$

(40), $\mathbf{B}_r{}^{ndr}(E_r,\mathbf{x},t) \in \partial I_r(\beta_r,(\mathbf{x},t))$,

$$\mathbf{H}_r{}^{nd}(E_r) = \frac{\partial \Psi_r}{\partial(\mathbf{grad}\beta_r)}(E_r) = \begin{vmatrix} k\mathbf{grad}\beta_1 \\ k\mathbf{grad}\beta_2 \end{vmatrix}.$$

The constitutive laws (34) to (38) give the constitutive laws for σ, $\mathbf{q}$ and for $\mathbf{B}_r$, $\mathbf{H}_r$, depending on $\mathbf{x}$,t, E_r and $\delta E_r = (\frac{d\varepsilon}{dt}, \frac{d\beta_1}{dt}, \frac{d\beta_2}{dt}, \mathbf{grad}\frac{d\beta_1}{dt}, \mathbf{grad}\frac{d\beta_2}{dt}, \mathbf{grad}\frac{1}{T})$ if there is dissipation,

(41), $\sigma = \sigma_r{}^d(\mathbf{x},t,E_r,\delta E_r,\chi) + \sigma_r{}^{nd}(E_r)$,

(42), $\mathbf{B}_r = \mathbf{B}_r{}^d(\mathbf{x},t,E_r,\delta E_r,\chi) + \mathbf{B}_r{}^{nd}(E_r) + \mathbf{B}_r{}^{ndr}(E_r,\mathbf{x},t)$,

(43), $\mathbf{H}_r = \mathbf{H}_r{}^d(\mathbf{x},t,E_r,\delta E_r,\chi) + \mathbf{H}_r{}^{nd}(E_r)$,

(44), $T\mathbf{q} = \mathbf{Q}_r{}^d(\mathbf{x},t,E_r,\delta E_r,\chi)$,

and

(45), $\beta_3 = (1-\beta_1-\beta_2)$.

The dissipative forces $\sigma_r{}^d$, $\mathbf{B}_r{}^d$, $\mathbf{H}_r{}^d$ and $\mathbf{Q}_r{}^d$ can be defined with the functions σ^d, $\mathbf{B}^d$, $\mathbf{H}^d$ and $\mathbf{Q}^d$, with

$$\mathbf{H}_r{}^d(\mathbf{x},t,E_r,\delta E_r,\chi) = \begin{vmatrix} H_1{}^d(\mathbf{x},t,E,\delta E,\chi)-H_3{}^d(\mathbf{x},t,E,\delta E,\chi) \\ H_2{}^d(\mathbf{x},t,E,\delta E,\chi)-H_3{}^d(\mathbf{x},t,E,\delta E,\chi) \end{vmatrix},$$

$$\mathbf{B}_r{}^d(\mathbf{x},t,E_r,\delta E_r,\chi) = \begin{vmatrix} B_1{}^d(\mathbf{x},t,E,\delta E,\chi)-B_3{}^d(\mathbf{x},t,E,\delta E,\chi) \\ B_2{}^d(\mathbf{x},t,E,\delta E,\chi)-B_3{}^d(\mathbf{x},t,E,\delta E,\chi) \end{vmatrix}, \text{ with } \beta_3 = (1-\beta_1-\beta_2),$$

and

$$\sigma_r{}^d(\mathbf{x},t,E_r,\delta E_r,\chi) = \sigma^d(\mathbf{x},t,E,\delta E,\chi) \text{ and } \mathbf{Q}_r{}^d(\mathbf{x},t,E_r,\delta E_r,\chi) = \mathbf{Q}^d(\mathbf{x},t,E,\delta E,\chi),$$

with $\beta_3 = (1-\beta_1-\beta_2)$,

or directly with functions $\sigma_r{}^d(\mathbf{x},t,E_r,\delta E_r,\chi)$, $\mathbf{B}_r{}^d(\mathbf{x},t,E_r,\delta E_r,\chi)$, $\mathbf{H}_r{}^d(\mathbf{x},t,E_r,\delta E_r,\chi)$ and $\mathbf{Q}_r{}^d(\mathbf{x},t,E_r,\delta E_r,\chi)$ which satisfy the inequality,

$\forall \mathbf{x},t$, $\forall E_r$, $\forall W_r = (f,c_r,g_r,\mathbf{p}) \in S \times \mathbf{R}^2 \times \mathbf{R}^{2\times 3} \times \mathbf{R}^3$, $\forall \chi$,

(14bis), $\sigma_r^d(x,t,E_r,W_r,\chi):f + \mathbf{B}_r^d(x,t,E_r,W_r,\chi)c_r + \mathbf{H}_r^d(x,t,E_r,W_r,\chi).g_r + \mathbf{Q}_r^d(x,t,E_r,W_r,\chi).\mathbf{p} \geq 0.$

The constitutive laws we have are (40) to (44). Let us one more time emphasis that the internal constraints (30) and (31) are part of the constitutive laws.

9.6. <u>The first non-dissipative model.</u>

A model is a set of partial differential equations able to describe the evolution of a structure i.e. able to compute the state E(x, t) knowing the initial situation at the initial time t = 0 and the external actions applied to the structure. In the case of a non dissipative model where **grad**β is not a state quantity, $E = (\varepsilon,\beta_1,\beta_2,\beta_3,T)$ or $E_r = (\varepsilon,\beta_1,\beta_2,T)$ and $T\underline{h} = \underline{I}$ or $T\underline{h}_r = I_r$. In fact we want to compute ($\mathbf{u}(\mathbf{x}, t),\beta_1(\mathbf{x}, t),\beta_2(\mathbf{x}, t),T(\mathbf{x}, t)$) where **u** is the small displacement. The equations are

(2), $\rho\gamma = \text{div}\sigma + \mathbf{f}$, in Ω,

(3), $\sigma.N = \mathbf{T}$, in $\partial\Omega$,

(5 quatro), $0 = -\mathbf{B}_r + \mathbf{A}_r$, in Ω,

where Ω is the domain occupied by the structure. In the sequel we assume the external action $\mathbf{A}_r$ to be equal to 0 ;

(7bis), $\dfrac{de_r}{dt} + \text{div}\mathbf{q} = r + \sigma{:}D(U) + \mathbf{B}_r.\dfrac{d\beta_r}{dt}$, in Ω,

(8), $-\mathbf{q}.N = \pi$, in $\partial\Omega$;

$$\sigma = \sigma^{nd}(E_r) = \frac{\partial\Psi_r}{\partial\varepsilon}(E_r),$$

$$\mathbf{B}_r^{nd}(E_r) = \frac{\partial\Psi_r}{\partial\beta_r}(E_r) = \mathbf{Y}_r(E_r) = \begin{vmatrix} \Psi_{1r}(E_r)-\Psi_{3r}(E_r) \\ \Psi_{2r}(E_r)-\Psi_{3r}(E_r) \end{vmatrix},$$

(40), $\mathbf{B}_r^{ndr}(E_r,\mathbf{x},t) \in \partial I_r(\beta_r,(\mathbf{x},t))$,

(46), $\mathbf{B}_r^{nd}(E_r)+\mathbf{B}_r^{ndr}(E_r,\mathbf{x},t) = \mathbf{B}_r = \mathbf{A}_r = 0$,

$$\mathbf{q} = -\lambda\,\mathbf{grad}T,$$

and

(45), $\beta_3 = (1-\beta_1-\beta_2)$;

$$e_r = \Psi_r - Ts_r.$$

This long list of equation gives a set of partial differential equations for the unknowns ($\mathbf{u},\beta_1,\beta_2,T$). It is completed with boundary conditions and initial conditions for the equations (2) and (7bis). This set of partial differential equations can be investigated with mathematics and numerical methods [8], [30], [31], [24], [33]. This model will be the basic model we study in the sequel.

9.7. The second non-dissipative model.
It is the previous model with a new interaction function,

$$\mathrm{Th}_r(E_r) = \frac{k}{2}(\mathbf{grad}\beta_1)^2+\frac{k}{2}(\mathbf{grad}\beta_2)^2+I_r(\beta_r),$$

The equations which are new or different from those of the previous paragrah are:

the equations of movement,

(5ter), $0 = \mathrm{div}\mathbf{H}_r - \mathbf{B}_r + \mathbf{A}_r$, in Ω,

(6ter), $\mathbf{H}_r.\mathbf{N} = \mathbf{a}_r$, in $\partial\Omega$;

the energy balance equation,

(7ter), $\frac{de_r}{dt} + \mathrm{div}\mathbf{q} = r + \sigma{:}D(U) + \mathbf{B}_r.\frac{d\beta_r}{dt} + \mathbf{H}_r.\mathbf{grad}\frac{d\beta_r}{dt}$, in Ω,

and the new constitutive law

(47), $\mathbf{H}_r = \mathbf{H}_r^{nd}(E_r) = \frac{\partial\Psi_r}{\partial(\mathbf{grad}\beta_r)}(E_r)$.

The list of equation is a long one! The partial differential equations are (2), (3), (5ter), (6ter) and (7ter), (8). The constitutive laws are (40), (41) with $\sigma_r^d = 0$, (42) with $\mathbf{B}_r^d = 0$, (43) with $\mathbf{H}_r^d = 0$, (44), (45) and (47).

9.8. An example of non-dissipative evolution.
Let us consider a unidimendional experiment and assume ε_{11} to be the only non zero deformation. Let us also assume that $\mathbf{grad}\beta_i = 0$ in the second model for the results to apply to both the two non-dissipative models. Let us focus on the stress σ_{11} as a function of ε_{11} when the temperature is fixed. From relation (40) we have

(48), $\sigma_{11} = K_{1111}\varepsilon_{11}+\tau_{11}(T)(\beta_2-\beta_1)$,

from relations (41), (42) and (43) we get,

(49), $-\mathbf{Y}_r(E_r) \in \partial I_r(\beta_r)$,

with

$$\mathbf{Y}_r(E_r) = \left| \begin{array}{c} -\tau_{11}(T)\varepsilon_{11} + \frac{l_a}{T_0}(T-T_0) \\ \tau_{11}(T)\varepsilon_{11} + \frac{l_a}{T_0}(T-T_0) \end{array} \right|.$$

The relation (49) means that the vector $-\mathbf{Y}_r(E_r)$ is normal to the triangle C_r at the point β_r (figure 2).

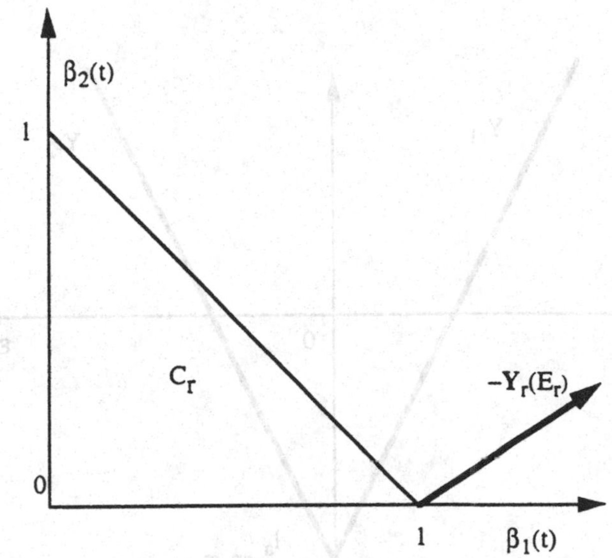

Figure 2
The vector $-Y_r(E_r)$ is normal to the triangle C_r.

9.8.1. Low temperature behaviour ($T < T_0$).

The temperature is fixed and low, $T < T_0$. We look for $\sigma_{11} = \sigma$, β_1 and β_2 as functions of $\varepsilon_{11} = \varepsilon$. The two components of $Y_r(E_r)$ $Y_{r1}(E_r) = -\tau_{11}(T)\varepsilon + \frac{l_a}{T_0}(T-T_0)$ and $Y_{r2}(E_r) = \tau_{11}(T)\varepsilon + \frac{l_a}{T_0}(T-T_0)$, are shown on figure 3.

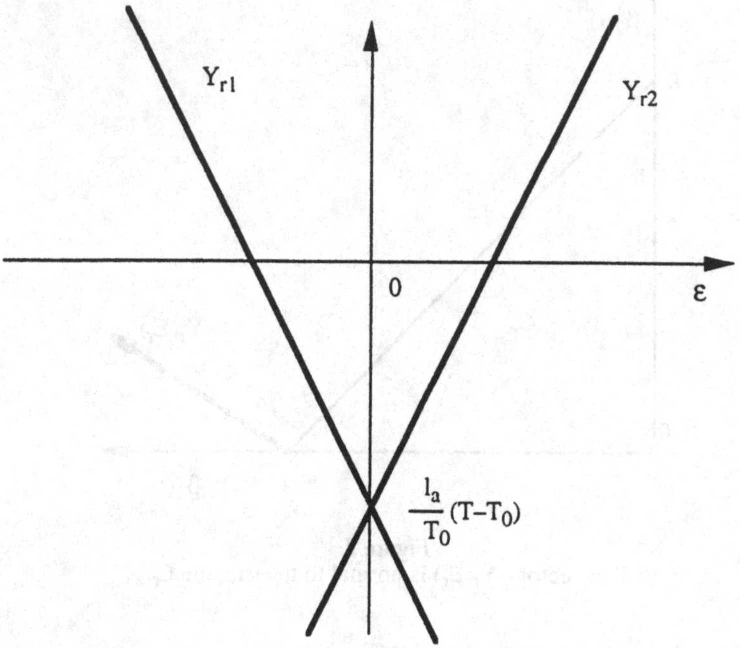

Figure 3
The components of $\mathbf{Y}_r(E_r)$ versus $\varepsilon = \varepsilon_{11}$ at low temperature.

When $\varepsilon = 0$, the two component of the vector $-\mathbf{Y}_r(E_r)$ are equal and positive. This vector can be normal to the triangle C_r only on the side AB (figure 4). Thus $\beta_3 = 0$ and we have a mixture of the two martensites. The stress σ can take any value of the segment $[\tau_{11}(T), -\tau_{11}(T)]$ (we have $\tau_{11}(T) = (T-T_c)\underline{\tau}_{11} > 0$ because we have assumed $\underline{\tau}_{11} < 0$).

When $\varepsilon > 0$, we have $-Y_{r1}(E_r) > 0$ and $-Y_{r1}(E_r) > -Y_{r2}(E_r)$. The only point where $-\mathbf{Y}_r(E_r)$ can be normal to the triangle C_r is the vertex A (figure 4).

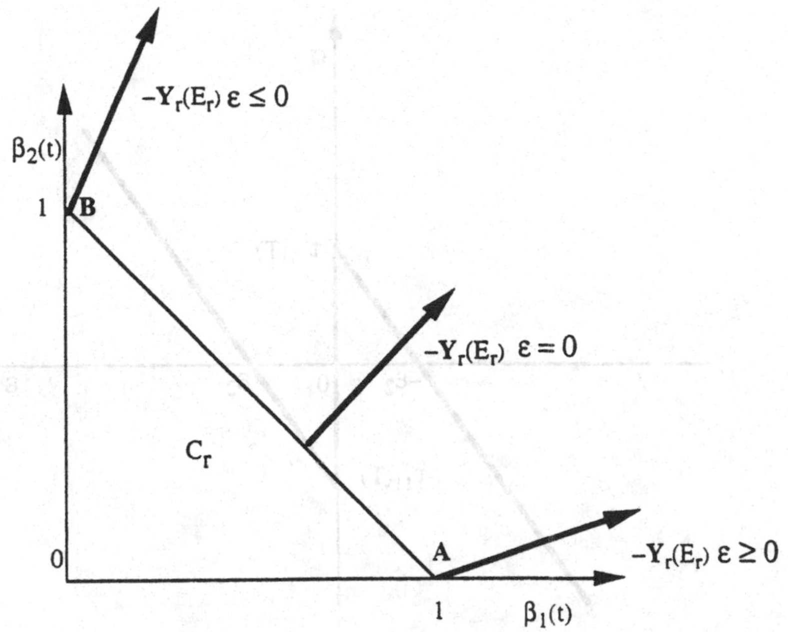

Figure 4
The vector $-Y_r(E_r)$ for different deformations ε at low temperature.

Thus $\beta_1 = 1$: there is only the martensite number one and $\sigma = K\varepsilon - \tau_{11}(T)$ (figure 5), (we let $K = K_{1111}$).

When $\varepsilon < 0$, we have $-Y_{r2}(E_r) > 0$ and $-Y_{r2}(E_r) > -Y_{r1}(E_r)$. The vector $-Y(E_r)$ is normal to the triangle C_r at the vertex B (figure 4). Thus $\beta_2 = 1$: there is only the martensite number two and $\sigma = K\varepsilon + \tau_{11}(T)$, (figure 5).

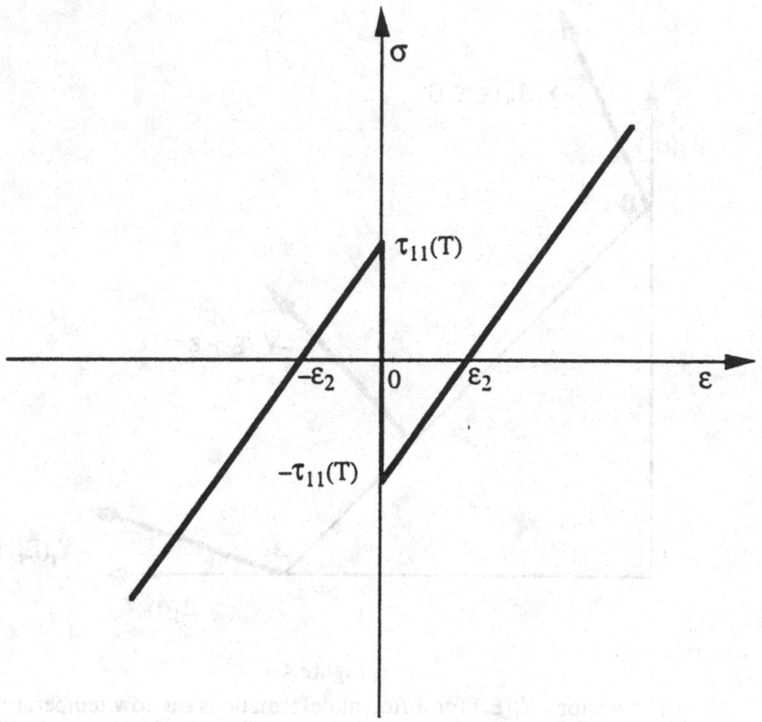

Figure 5
The non-dissipative constitutive low at low temperature.

We get some of the properties of shape memory alloys : at low temperature there is no austensite but mixtures of martensites ; there is a softening of the behaviour when going from compression to tension. Of course the behaviour at the origin is not the actual one but it has some of its properties.

9.8.2. Medium temperature behaviour ($T_0 < T < T_c$).
The temperature is fixed and satisfies $T_0 \leq T \leq T_c$. The two components of the vector $\mathbf{Y}_r(E_r)$ are shown on figure 6.

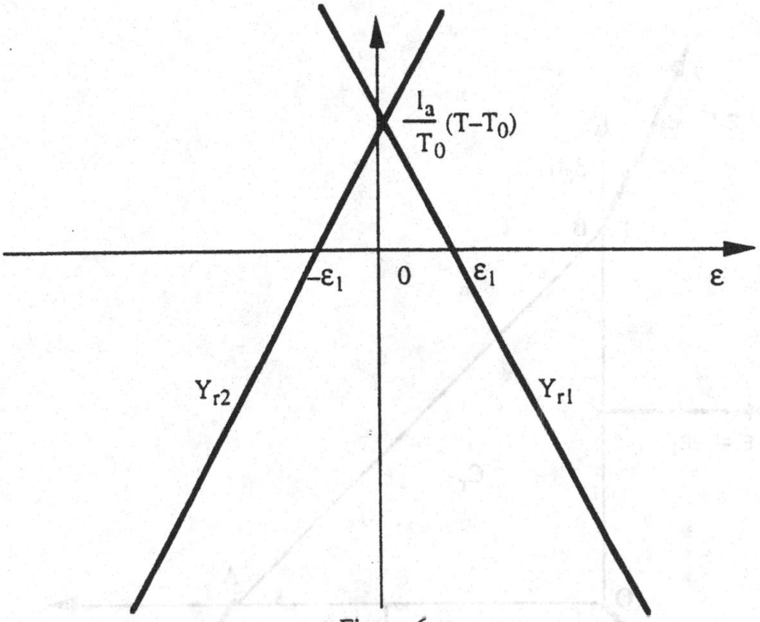

Figure 6
The components of $\mathbf{Y}_r(E_r)$ at medium temperature.

When $\varepsilon = 0$ the two components of $\mathbf{Y}_r(E_r)$ are equal and negative. The vector $\mathbf{Y}_r(E_r)$ can be normal to the triangle C_r only at the vertex 0 (figure 7).

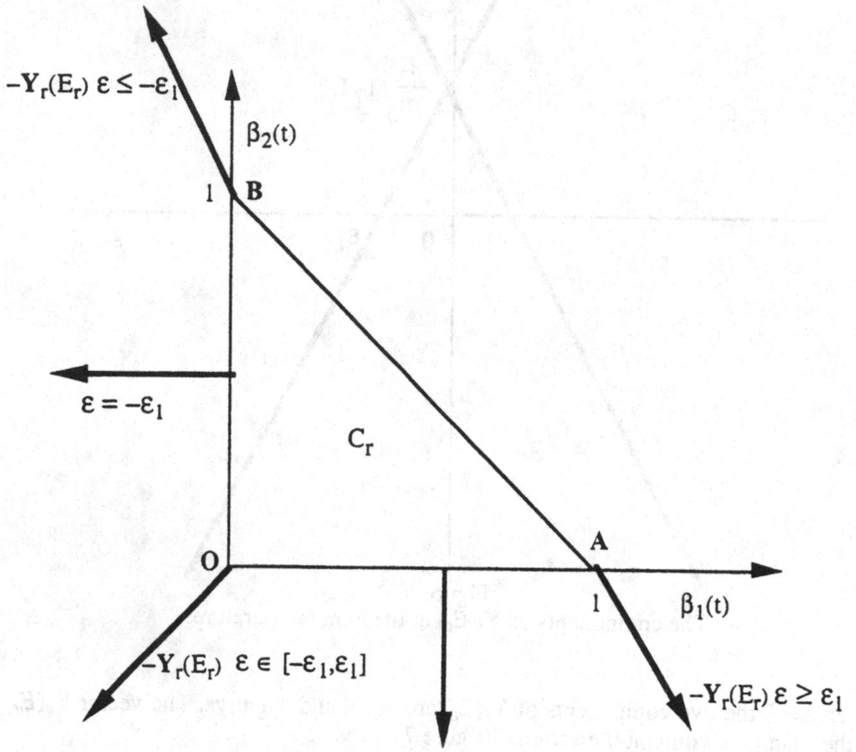

Figure 7
The vector $-Y_r(E_r)$ for different deformations ε at medium temperature.

Thus $\beta_3 = 1$ and $\beta_1 = \beta_2 = 0$, there is only austenite. Relation (27) gives $\sigma = 0$ for $\varepsilon = 0$. When $\varepsilon \neq 0$, the vector $Y_{r1}(E_r)$ is normal to C_r at vertex 0 if its two components are negative i.e. for

$$-\frac{l_a(T-T_0)}{T_0\tau_{11}(T)} \le \varepsilon \le \frac{l_a(T-T_0)}{T_0\tau_{11}(T)} = \varepsilon_1.$$

Thus for $\varepsilon \in]-\varepsilon_1,\varepsilon_1[$, there is only austenite, $\beta_3 = 1$, and from (48), $\sigma = K\varepsilon$ (figure 8).

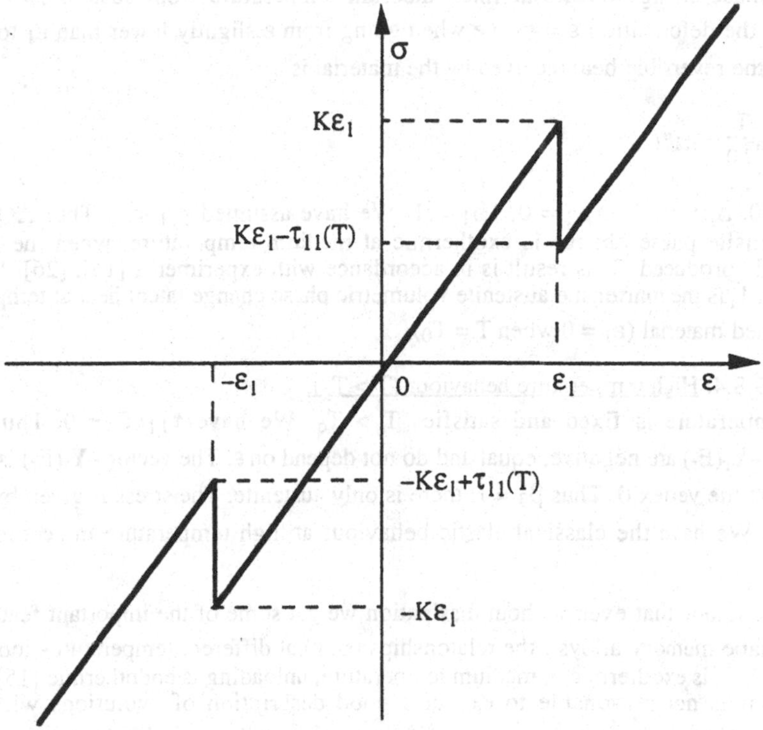

Figure 8
The constitutive low at medium temperature.

For $\varepsilon = \varepsilon_1$, we have $Y_{r1}(E_r) = 0$. The vector $-Y_r(E_r)$ is normal to the triangle C_r on the side OA (figure 7) and $\beta_1+\beta_3 = 1$: we have a mixture of austenite and martensite number one. The stress σ can take any value of the segment $[K\varepsilon_1, K\varepsilon_1-\tau_{11}(T)]$ (figure 8).

For $\varepsilon > \varepsilon_1$, we have $-Y_{r1}(E_r) > 0$ and $-Y_{r2}(E_r) < 0$. The vector $-Y_r(E_r)$ is normal to the triangle C_r at the vertex A (figure 7) and $\beta_1 = 1$, there is only the martensite number one and the stress is $\sigma = K\varepsilon-\tau_{11}(T)$.

The increase of deformation produces the martensite-austenite phase change. We have the same result when decreasing the deformation: phase change from austenite to martensite number two occurs at $\varepsilon = -\varepsilon_1$. When $\varepsilon < -\varepsilon_1$ there is only martensite number two, $\beta_2 = 1$, and the stress is $\sigma = K\varepsilon+\tau_{11}(T)$, (figure 8).

9.8.3. <u>Austenite martensite phase change latent heat.</u>

Let us compute

$$Tds = -Td(\frac{\partial\Psi}{\partial T}) = CdT+l_a\frac{T}{T_0}\,d\beta_3-\underline{\tau}^T{:}\varepsilon(d\beta_2-d\beta_1)-(\beta_2-\beta_1)\underline{\tau}^T{:}d\varepsilon$$

$$= Tds_r = CdT-(\underline{\tau}^T{:}\varepsilon+l_a\frac{T}{T_0})d\beta_2+(\underline{\tau}^T{:}\varepsilon-l_a\frac{T}{T_0})d\beta_1-(\beta_2-\beta_1)\underline{\tau}^T{:}d\varepsilon.$$

When phase-change occurs at fixed medium temperature from austenite to martensite number one, at the deformation $\varepsilon = \varepsilon_l$, i.e when going from ε slightly lower than ε_l to ε slitghtly greater than ε_l the reversible heat received by the material is

$$\Delta Q = -l_a \frac{T}{T_0} + \underline{\tau}_{11}\varepsilon_l,$$

because $\Delta T = 0$, $\Delta\beta_3 = -1$, $\Delta\beta_2 = 0$, $\Delta\beta_1 = 1$. We have assumed $\underline{\tau}_{11} < 0$. Thus $\Delta Q < 0$. The austenite-martensite phase change is exothermic at medium temperature: when the material is deformed heat is produced. This result is in accordance with experiments [15], [26]. Let us note that the quantity l_a is the martensite austenite volumetric phase change latent heat at temperature T_0 of the undeformed material ($\varepsilon_l = 0$ when $T = T_0$).

9.8.4. High temperature behaviour ($T > T_c$).

The temperature is fixed and satisfies $T > T_c$. We have $\tau_{11}(T) = 0$. Thus the two components of $-Y_r(E_r)$ are negative, equal and do not depend on ε. The vector $-Y_r(E_r)$ is normal to the triangle C_r at the vertex 0. Thus $\beta_3 = 1$: there is only austenite. The stress is given by (48), $\sigma = K\varepsilon$ (figure 26). We have the classical elastic behaviour at high temperature in accordance with experiments.

Let us conclude that even without dissipation we get some of the important features of the behaviour of shape memory alloys : the relatonships (σ, ε) at different temperatures looks like the actual ones; loading is exothermic at medium temperature, unloading is endothermic [15].

Even if it is not reasonable to expect a good description of evolutions with this non dissipative constitutive law, let us consider a deformed unloaded material at low temperature ($\varepsilon = \varepsilon_2 = \frac{\tau_{11}(T)}{K}$, $\sigma = 0$, figure 5) and heat it to high temperature : the material goes back to the undeformed state ($\varepsilon = 0$, $\sigma = 0$). This is one kind of shape memory!

The results will be much more better with an educated alloy.

9.9. The dissipation. The pseudo-potential of dissipation.

From experiments it is known that the behaviour of shape memory alloys depends on time, i.e the behaviour is dissipative. It results that the mechanical part of the pseudo-potential of dissipation Φ is not zero.

Again to deal only with the main features of the behaviour, we assume there is only dissipation with respect to the $\frac{\partial\beta_i}{\partial t}$. Because relation (45) implies

$$\frac{\partial\beta_3}{\partial t} = -\frac{\partial\beta_1}{\partial t} - \frac{\partial\beta_2}{\partial t},$$

we can eliminate the velocity $\frac{\partial\beta_3}{\partial t}$ in the constitutive laws. As we have already mentionned, we can also define directly the dissipative internal forces. We choose this way and define a pseudo-potential of dissipation,

$$(50),\quad \Phi_r(T,\frac{\partial\beta_1}{\partial t},\frac{\partial\beta_2}{\partial t},\mathbf{grad}\frac{1}{T}) = c\sqrt{\{(\frac{\partial\beta_1}{\partial t})^2+(\frac{\partial\beta_2}{\partial t})^2\}}+k\{(\frac{\partial\beta_1}{\partial t})^2+(\frac{\partial\beta_2}{\partial t})^2\}+\frac{\lambda T^3}{2}(\mathbf{grad}\frac{1}{T})^2,$$

with $c \geq 0$ and $k \geq 0$. Of course this expression is induced by experimental results: the first term is related to the permanent deformations exhibited by experiments; the second term is related to the viscous aspect. It has a smoothing effect. We have also

$$\Phi_r(T,\frac{\partial\beta_1}{\partial t},\frac{\partial\beta_2}{\partial t},\mathbf{grad}\frac{1}{T}) = c\left\|\frac{\partial\beta_r}{\partial t}\right\| + k(\left\|\frac{\partial\beta_r}{\partial t}\right\|)^2+\frac{\lambda T^3}{2}(\mathbf{grad}\frac{1}{T})^2,$$

where the euclidian norm is

$$\left\|\frac{\partial\beta_r}{\partial t}\right\| = \sqrt{\{(\frac{\partial\beta_1}{\partial t})^2+(\frac{\partial\beta_2}{\partial t})^2\}}.$$

9.10. <u>The dissipative constitutive laws.</u>

Following paragraph 9.5 and assuming there is no quantity , we define the dissipative forces,

$$\sigma_r^d(x,t,E_r,\delta E_r) = 0,$$

$$\mathbf{H}_r^d(x,t,E_r,\delta E_r) = 0,$$

$$\mathbf{B}_r^d(\delta E_r) \in \partial\Phi_r(\frac{\partial\beta_1}{\partial t},\frac{\partial\beta_2}{\partial t}),$$

where the subdifferential of the non-smooth function Φ_r is

$$\partial\Phi_r(x,t,\frac{\partial\beta_1}{\partial t},\frac{\partial\beta_2}{\partial t}) = c\frac{\frac{\partial\beta_r}{\partial t}}{\left\|\frac{\partial\beta_r}{\partial t}\right\|}+k\frac{\partial\beta_r}{\partial t},\ \text{if}\ \left\|\frac{\partial\beta_r}{\partial t}\right\| \neq 0,$$

$$\partial\Phi_r(x,t,0,0) \in S\ ,\ \text{if}\ \left\|\frac{\partial\beta_r}{\partial t}\right\| = 0,\text{with}\ S=\{S\in\mathbf{R}^2;\ \|S\|\leq c\}.$$

The constitutive laws are then

$$(51),\quad \sigma = \sigma_r^{nd}(E_r),$$

$$(41),\quad \mathbf{B}_r^{nd}(E_r) = \frac{\partial\Psi_r}{\partial\beta_r}(E_r) = \mathbf{Y}_r(E_r) = \begin{vmatrix}\Psi_{1r}(E_r)-\Psi_{3r}(E_r)\\ \Psi_{2r}(E_r)-\Psi_{3r}(E_r)\end{vmatrix},$$

$$(42),\quad \mathbf{B}_r^{ndr}(E_r,x,t) \in \partial I_r(\beta_r(x,t)),$$

$$(52),\quad \mathbf{B}_r = \mathbf{B}_r^{nd}(E_r)+\mathbf{B}_r^{ndr}(E_r,x,t) + \mathbf{B}_r^d(x,t,\frac{\partial\beta_1}{\partial t},\frac{\partial\beta_2}{\partial t}),$$

(53), $\mathbf{H}_r = \mathbf{H}_r^{nd}(E_r)$,

(18), $\mathbf{Tq} = \mathbf{Q}^d(E_r,\delta E_r) = \lambda T^3 \mathbf{grad}\frac{1}{T}$ or $\mathbf{q} = -\lambda \mathbf{grad}T$.

9.11. The dissipative behaviour.

Although the dissipation is more interesting to describe evolutions of the material, let us briefly look for the equilibrium states, i.e find the (σ,ε) and β_r such that $\frac{\partial \beta_r}{\partial t} = 0$ in relations (52) and (5bis). We have already assumed that the external action $\mathbf{A}_r$ is equal to 0. For the sake of simplicity let us also assume that in the sequel $\mathbf{grad}\beta_i = 0$. So the results apply at the two models taking or not taking into account the $\mathbf{grad}\beta_i$. It results from our assumptions ($\mathbf{A}_r = 0$ and $\mathbf{grad}\beta_i = 0$) and from the movement equations (5) or (5bis) that

(52bis), $$0 = \mathbf{B}_r^{nd}(E_r) + \mathbf{B}_r^{ndr}(E,\mathbf{x},t) + \mathbf{B}_r^d(\mathbf{x},t,\frac{\partial \beta_1}{\partial t},\frac{\partial \beta_2}{\partial t}).$$

9.11.1. Equilibrium at low temperature.

Because we have $\frac{\partial \beta_r}{\partial t} = 0$, $\mathbf{B}_r^d \in S$. The constitutive laws (41), (42) and the relation (52bis) give

$$\mathbf{Y}_r + \mathbf{S}_r + \mathbf{R}_r = 0, \text{ with } \mathbf{S}_r \in S,\ \mathbf{R}_r \in \partial I(\beta_r).$$

This relation means that the vector $-(\mathbf{Y}_r + \mathbf{S}_r)$ is normal to the triangle C_r.

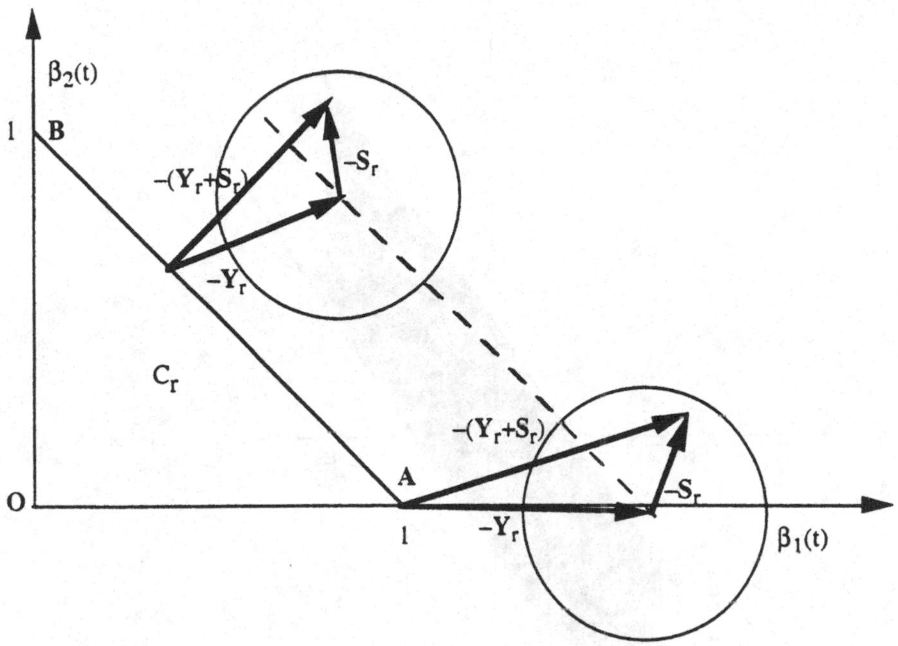

Figure 9
The vector - $(\mathbf{Y}_r+\mathbf{S}_r)$ is normal to the triangle C_r.

Let $\varepsilon = 0$, the vectors $-(\mathbf{Y}_r+\mathbf{S}_r)$ (figure 9) have positive components (we have assumed $\frac{l_a}{T_0}(T_0-T) > c$). Thus the only possibility for $-(\mathbf{Y}_r+\mathbf{S}_r)$ to be normal to the triangle is to be normal on the side AB. Thus $\beta_1+\beta_2 = 1$: we have a mixture of the two martensites.

Let us increase ε, the vector $-\mathbf{Y}_r$ leans slightly on the side AB but because of $\mathbf{S}_r$ it is still possible for the vector $-(\mathbf{Y}_r+\mathbf{S}_r)$ to be normal to the triangle C_r on the side AB. We still have a mixture of the two martensites: $\beta_1+\beta_2 = 1$. The stress is $\sigma \in K\varepsilon+[\tau_{11}(T), -\tau_{11}(T)]$, (figure 10).

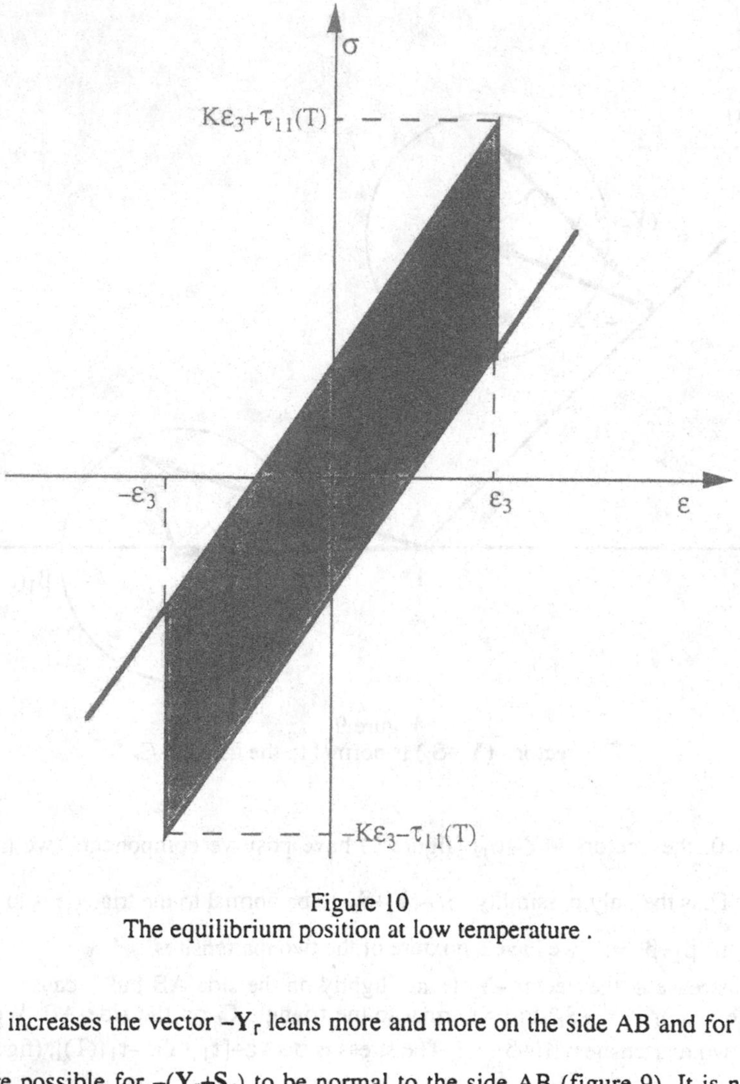

Figure 10
The equilibrium position at low temperature .

When ε increases the vector $-\mathbf{Y}_r$ leans more and more on the side AB and for $\varepsilon > \dfrac{c}{\tau_{11}(T)} =$ ε_3 it is no more possible for $-(\mathbf{Y}_r+\mathbf{S}_r)$ to be normal to the side AB (figure 9). It is normal to the triangle at the vertex A: $\beta_1 = 1$, there is only the martensite number one. The stress is (figure 10), $\sigma = K\varepsilon_3 - \tau_{11}(T)$.

Note. We have assumed $\varepsilon_3 > \varepsilon_1$ or $cK = cK_{1111} > \tau_{11}^2(T)$. The results are slighty different when this assuption is not satisfied.

We have the symmetric result for ε negative (figure 10). The effect of dissipation is to increase the domain (σ,ε) where equilibrium is possible (figure 10).

9.11.2. Equilibrium at medium temperature.
The effect of the set S is the same, it allows the vector $-(\mathbf{Y}_r+\mathbf{S}_r)$ to be normal to the triangle

C_r on the side OA not only for $\varepsilon = \varepsilon_l$ but also for a small interval $[\varepsilon_4,\varepsilon_5]$ around ε_l (figures 11 and 12).

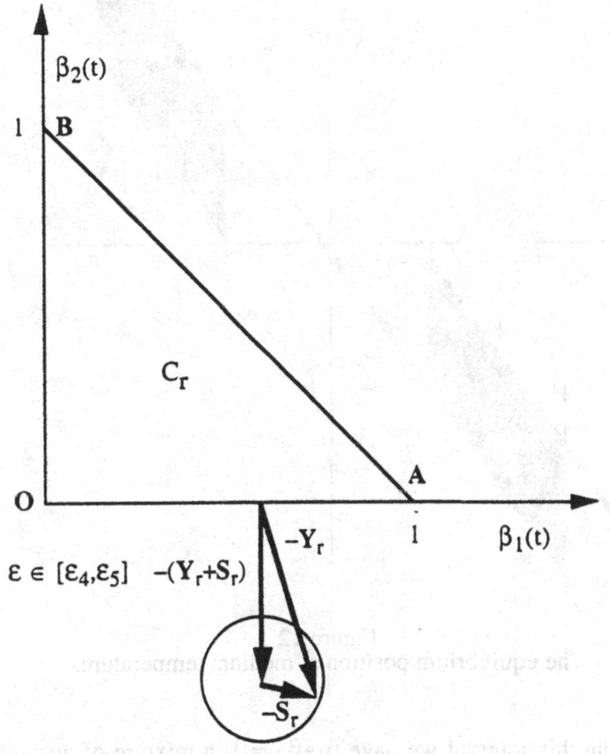

Figure 11
The vector - (Y_r+S_r) is normal to the triangle C_r.

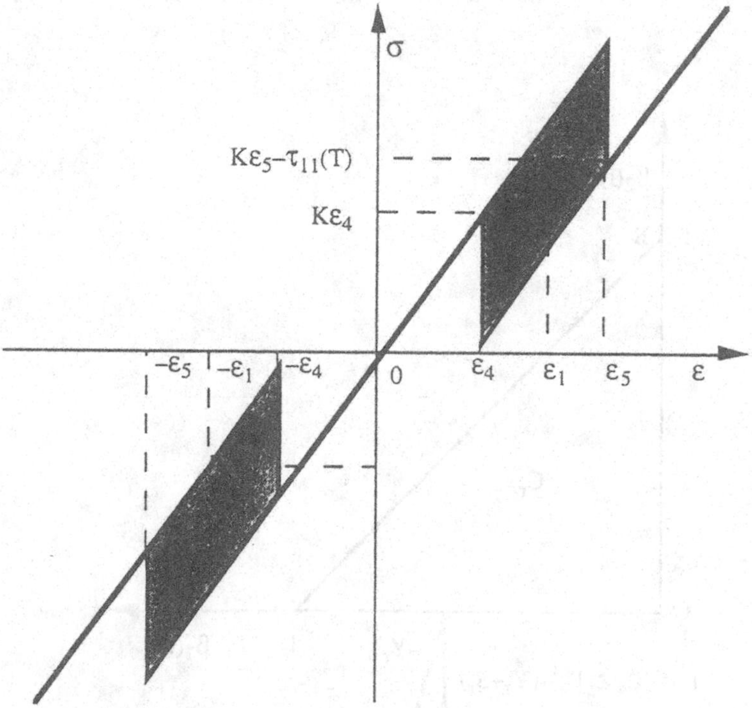

Figure 12
The equilibrium position at medium temperature.

When ε is within this interval we have $\beta_1+\beta_3 = 1$: a mixture of austenite and martensite number one. Thus the stress is $\sigma \in K\varepsilon+[-\tau_{11}(T),0]$.

An analogous result is obtained for ε negative (figure 12).

9.11.3. Equilibrium at high temperature.

The components of $-(Y_r+S_r)$ remain negative (we have assume $T > T_c$ and T large enough). Thus the vector $-(Y_r+S_r)$ is normal to the triangle C_r only at the vertex 0, (figure 13).

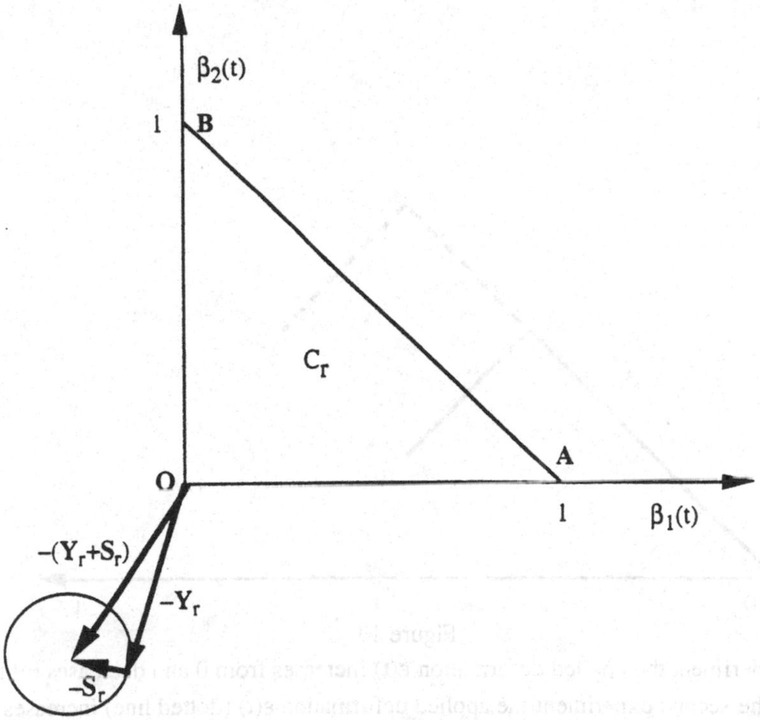

Figure 13
The vector - (Y_r+S_r) normal to the triangle C_r at high temperature.

We have only austenite, $\beta_3 = 1$, and $\sigma = K\varepsilon$ (figure 26). The equilibrium position are not modified at high temperature by the dissipation.

9.12. Evolution of a shape memory alloy.

The dissipative terms of the constitutive law (52) or of the relation (52bis) involve derivatives with respect to the time. Thus they are differential equations. The natural problem to look at is the evolution of a material submitted to time dependant external actions. In this paragraph we study unidimensional experiments. We choose to apply the deformation $\varepsilon(t)$ because it is more easy to inverstigate the structure of the equations and exhibit the hysteretical properties of shape memory alloys.

To be specific we look at three experiments. The two first are at fixed low temperature, the last one at fixed high temperature.

In the first experiment, the deformation $\varepsilon(t)$ starting from 0 increases then decreases till the stress σ is zero (figure 14).

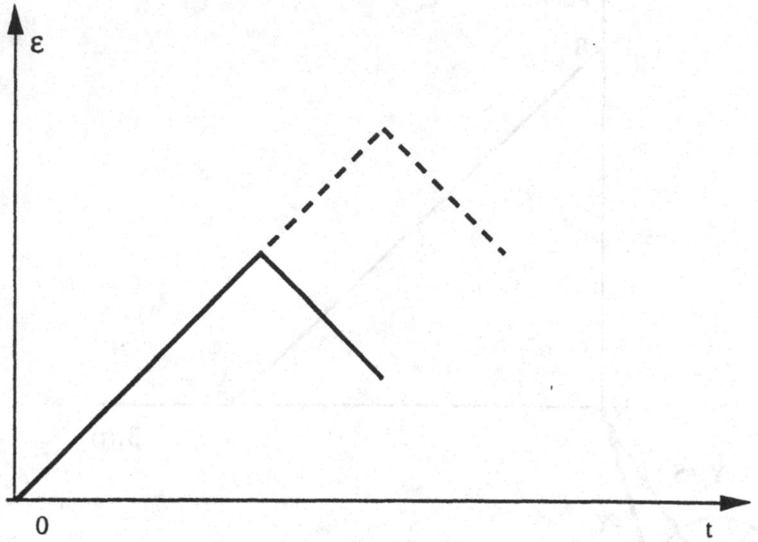

Figure 14

In the first experiment the applied deformation $\varepsilon(t)$ increases from 0 and decreases till the stress $\sigma(t)$ is zero. In the second experiment the applied deformation $\varepsilon(t)$ (dotted line) increases to a larger value, then decreases till the stress $\sigma(t)$ is zero.

In the second experiment, the deformation increases to a larger value then decreases again to a value such that the stress is zero (figure 14). In the last experiment, the deformation starting from 0 increases.

In all experiments the initial mixture is made of the two martensites with equal volume fractions $\beta_1(0) = \beta_2(0) = \frac{1}{2}$. The point which represents this mixture in the plane (β_1, β_2) is $\beta_r{}^0 = (\frac{1}{2}, \frac{1}{2})$.

9.12.1. <u>First experiment at low temperature ($T < T_0$)</u>.

The point $\beta_r{}^0$ is an equilibrium position when $\varepsilon = 0$. The deformation $\varepsilon(t)$ increases from 0 as shown in figure 14. The vector $-\mathbf{Y}_r$ which at $t = 0$ is normal to the line AB leans progressively towards it (figure 15).

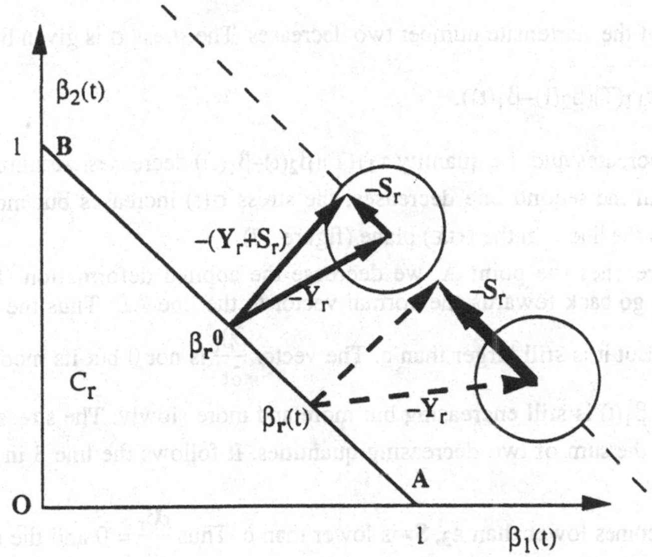

Figure 15
The vector - $(\mathbf{Y_r}+\mathbf{S_r})$ is normal to the triangle C_r.

When ε is not too large, $\varepsilon < \varepsilon_3 = \dfrac{c}{\tau_{11}(T)}$, it is possible to find $\mathbf{S_r} \in S$ and $\mathbf{R_r} \in \partial I_r(\beta_r)$ such that (52bis) is satisfied (figure 15). Thus we have $\dfrac{\partial \beta_r}{\partial t} = 0$. In the plane (σ, ε), (figure 17), this part of the evolution correspond to the line 1.

Note. The point $\beta_r(t)$ cannot go inside C_r because when $\beta_r(t)$ is inside $\mathbf{R_r} = 0$, $(\partial I_r(\beta_r) = \{0\})$. It would result from (52bis) that

$$-(\mathbf{Y_r}+\mathbf{S_r}) = 0,$$

which means that $\mathbf{S_r}$ and $\dfrac{\partial \beta_r}{\partial t}$ are directed outwards which is impossible because $\dfrac{\partial \beta_r}{\partial t}$ is directed inwards!

When $\varepsilon(t)$ goes beyong ε_3 it is no more possible to satisfy (52bis) with $\dfrac{\partial \beta_r}{\partial t} = 0$. Thus $\dfrac{\partial \beta_r}{\partial t}$ becomes non zero and is tangent to the triangle C_r (figure 15) with

$$\mathbf{S_r} = c\mathbf{t_r} + k\frac{\partial \beta_r}{\partial t},$$

where $\mathbf{t_r}$ is the unit vector $\dfrac{\mathbf{AB}}{|\mathbf{AB}|}$. The volume fraction of the martensite number one increases while the volume fraction of the martensite number two decreases. The stress σ is given by relation (48),

$$\sigma(t) = K\varepsilon(t)+\tau_{11}(T)(\beta_2(t)-\beta_1(t)).$$

The quantity $K\varepsilon(t)$ increases and the quantity $\tau_{11}(T)(\beta_2(t)-\beta_1(t))$ decreases. Assuming the first one to increase more than the second one decreases, the stress $\sigma(t)$ increases but more slowly than previously. It follows the line 2 in the (σ,ε) plane (figure 17).

Before $\beta_r(t)$ reaches the point A, we decrease the applied deformation (figure 14). The vector $-\mathbf{Y}_r$ begins to go back towards the normal vector to the line AB. Thus the modulus of the vector $\mathbf{S}_r$ decreases. But it is still larger than c. The vector $\frac{\partial\beta_r}{\partial t}$ is not 0 but its modulus decreases. The volume fraction $\beta_1(t)$ is still encreasing but more and more slowly. The stress $\sigma(t)$ decreases quickly because it is the sum of two decreasing quantities. It follows the line 3 in the (σ,ε) plane (figure 17).

When $\varepsilon(t)$ becomes lower than ε_3, $\mathbf{S}_r$ is lower than c. Thus $\frac{\partial\beta_r}{\partial t} = 0$ and the composition of the mixture β_r^1 remains constant. The value of the stress is

$$\sigma(t) = K\varepsilon(t)+\tau_{11}(T)(\beta^1_2-\beta^1_1).$$

It is shown in figure 17 on the line 4. The stress is zero for (figure 17),

$$\varepsilon_0 = \frac{-\tau_{11}(T)(\beta^1_2-\beta^1_1)}{K}.$$

The material remains in this position if no load is applied.

9.12.2. <u>Second experiment at low temperature ($T < T_0$).</u>

The deformation applied at the beginning is the one applied in the first experiment. The deformation $\varepsilon(t)$ is increased till the point $\beta_r(t)$ reaches the vertex **A**. At this point $\partial I_r(\mathbf{A})$ is a larger set and is possible to find $\mathbf{R}_r \in \partial I_r(\beta_r) = \partial I_r(\mathbf{A})$ and $\mathbf{S}_r \in S$ such that the vector $-(\mathbf{Y}_r+\mathbf{S}_r)$ is normal to the triangle (figure 16),

$$-(\mathbf{Y}_r+\mathbf{S}_r) = \mathbf{R}_r \in \partial I_r(\beta_r).$$

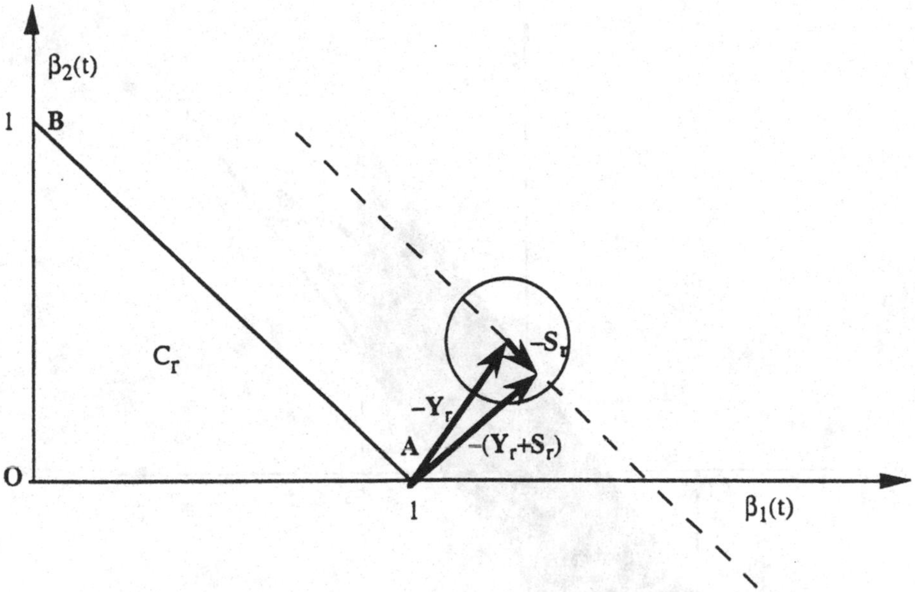

Figure 16
The vector - (Y_r+S_r) is normal to the triangle C_r at low temperature.

Thus we have $\frac{\partial \beta_r}{\partial t} = 0$. The stress σ versus the deformation ε before the point $\beta_r(t)$ reaches the vertex **A** is shown in figure 17 on the line 5.

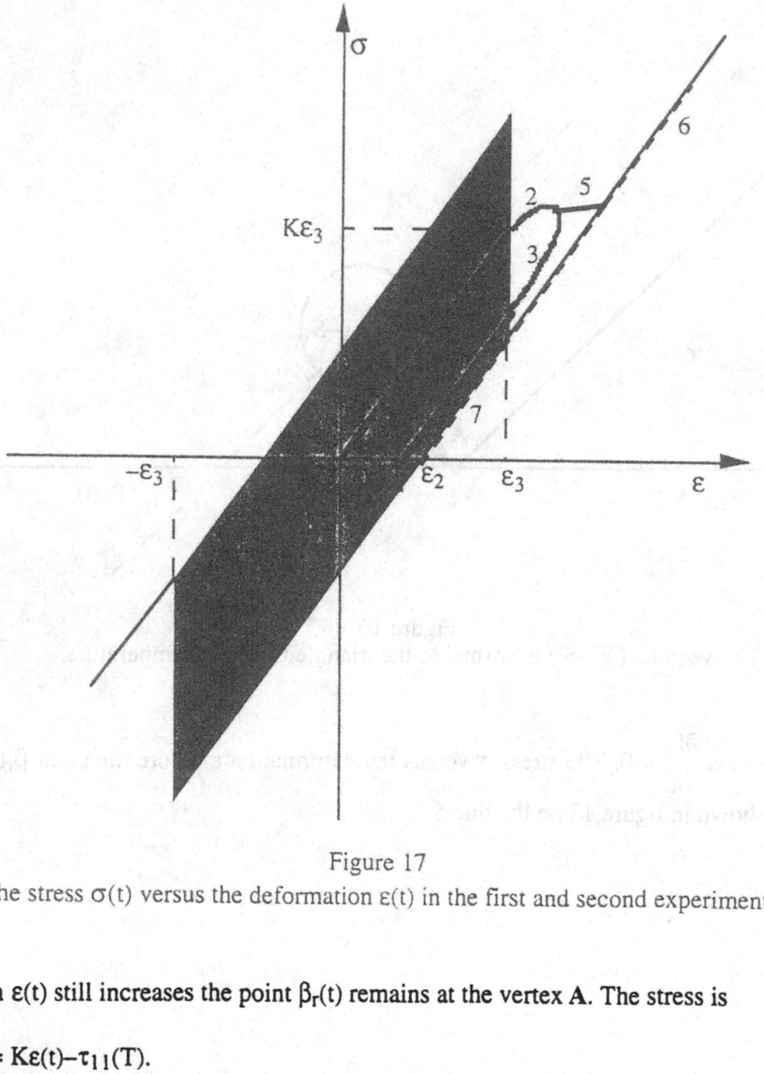

Figure 17
The stress $\sigma(t)$ versus the deformation $\varepsilon(t)$ in the first and second experiment.

When $\varepsilon(t)$ still increases the point $\beta_r(t)$ remains at the vertex **A**. The stress is

(54), $\sigma(t) = K\varepsilon(t) - \tau_{11}(T)$.

It is shown in figure 17 on the line 6. If we let $\varepsilon(t)$ decrease, we still have (52bis) with $S_r = 0$. Thus $\frac{\partial \beta_r}{\partial t} = 0$ and the stress is given by (54). It is zero for (figure 17),

$$\varepsilon = \varepsilon_2 = \frac{\tau_{11}(T)}{K}.$$

The stress σ is shown on the line 7 of figure 17. The material remains in the position ($\beta_1 = 1, \beta_2 = \beta_3 = 0, \sigma = 0, \varepsilon = \varepsilon_2$) if no load is applied.

Note. It is clear that any point of the segment $[-\varepsilon_2, \varepsilon_2]$ can be reached . We have assumed that $\varepsilon_2 = \frac{\tau_{11}(T)}{K} < \frac{c}{\tau_{11}(T)}$ or $\tau_{11}{}^2(T) < cK$. When it is not true the results are slightly different.

An other evolution at medium temperature is shown in figure 18 exhibiting the classical features of hysteresis. The applied deformations starts from 0, increases and comes back to 0. The point $(\sigma(t), \varepsilon(t))$ in figure 1! follows the path 1, 2, 3, 4, 5.

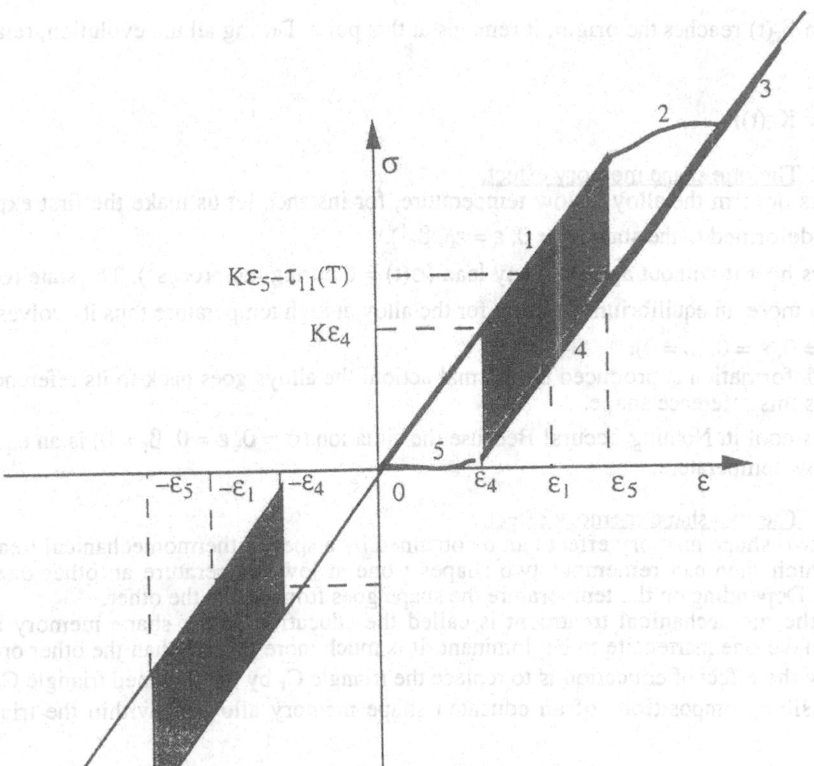

Figure 18

The stress $\sigma(t)$ versus the deformation $\varepsilon(t)$ in an experiment at medium temperature.

9.12.3. Third experiment at high temperature ($T > T_c$).

Let us repeat the first or second experiment at high temperature. The point $\beta_r{}^0$ is not an equilibrium point for $\beta_r(t)$. Thus the point $\beta_r(t)$ goes from $\beta_r{}^0$ to 0 on the segment $[\beta_r{}^0,0]$ following the equation $\mathbf{Y}_r+\mathbf{S}_r = 0$ or

$$\left\| \begin{matrix} \frac{l_a}{T_0}(T-T_0) \\ \frac{l_a}{T_0}(T-T_0) \end{matrix} \right\| + \left\| \begin{matrix} -\frac{c}{\sqrt{2}} + k\frac{\partial \beta_1}{\partial t} \\ -\frac{c}{\sqrt{2}} + k\frac{\partial \beta_1}{\partial t} \end{matrix} \right\| = 0.$$

Note. We have assumed T large enough for $\frac{l_a}{T_0}(T-T_0) - \frac{c}{\sqrt{2}} > 0$.

When $\beta_r(t)$ reaches the origin, it remains at this point. During all the evolution, relation (48) shows that

$$\sigma(t) = K\varepsilon(t).$$

9.13. The one shape memory effect.

Let us deform the alloy at low temperature, for instance let us make the first experiment. The alloy is deformed to the state ($\sigma = 0, \varepsilon = \varepsilon_0, \beta_r^1$).

Let us heat it without applying any load ($\sigma(t) = 0$ during the process). The state ($\sigma = 0, \varepsilon = \varepsilon_0, \beta_r^1$) is no more an equilibrium position for the alloy at high temperature thus it evolves towards the state ($\sigma = 0, \varepsilon = 0, \beta_r = 0$).

The deformation is produced by thermal action: the alloys goes back to its reference shape. It remembers this reference shape.

Let us cool it. Nothing occurs! Because the situation ($\sigma = 0, \varepsilon = 0, \beta_r = 0$) is an equilibrium position at low temperature.

9.14. The two shape memory effect.

The two shape memory effect can be obtained by a special thermomechanical treatment of the alloy which then can remember two shapes : one at low temperature an other one at high temperature. Depending on the temperature the shape goes from one to the other.

The thermomechanical treatment is called the education of the shape memory alloy. Its effect is to make one martensite to be dominant. It is much more present than the other one. In our point of view the effect of education is to replace the triangle C_r by the flattened triangle C_r^e (figure 19): the possible compositions of an educated shape memory alloy are within the triangle C_r^e (figure 19).

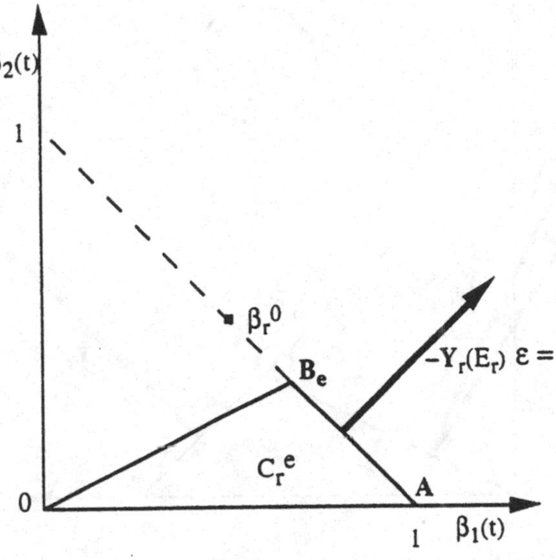

Figure 19
The educated shape memory triangle C_r^e. The possible compositions of the educated shape memory alloy are within the triangle C_r^e.

It is obvious that the martensite number one is much more present than the martensite number two. We assume that the point $\beta_r^0 = (\frac{1}{2},\frac{1}{2})$ does not belong to C_r^e.

9.14.1. Equilibrium states at low temperature ($T < T_0$).

In the plane (σ,ε) the possible equilibrium positions are shown in figure 21.

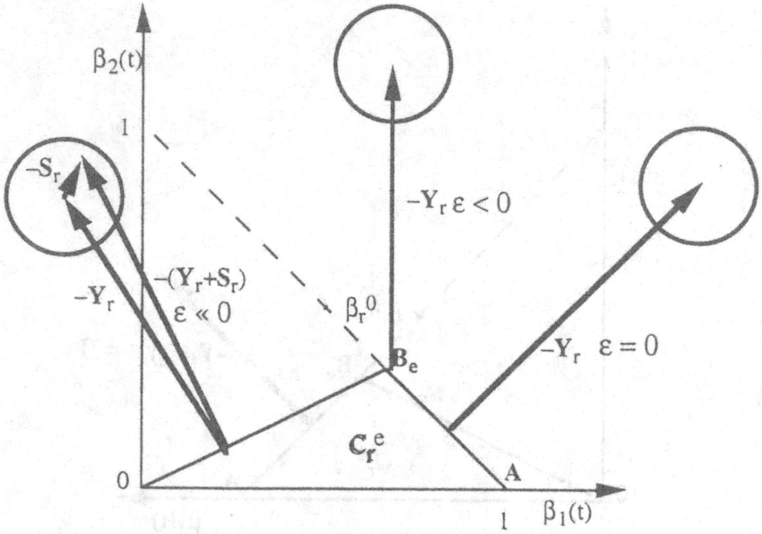

Figure 20
The vector - (Y_r+S_r) is normal to the triangle C_r^e.

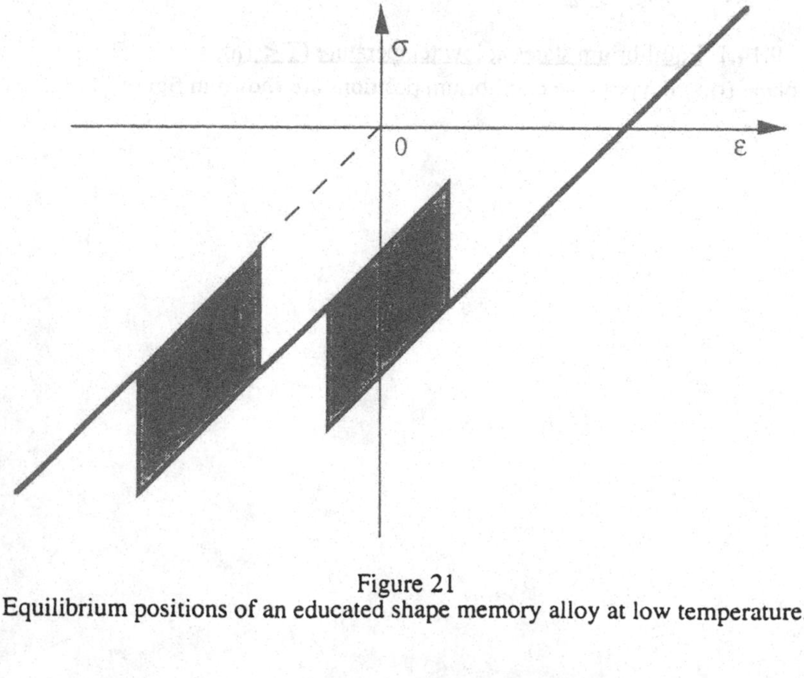

Figure 21
Equilibrium positions of an educated shape memory alloy at low temperature.

They result from equation (52bis) or

$$Y_r+S_r+R_r=0,$$

with $S_r\in S$ and $R_r\in\partial I_r^e(\beta_r)$, where I_r^e is the indicator function of the triangle C_r^e. Let us remind that equation (52bis) means that the vector $-(Y_r+S_r)$ is normal to the triangle C_r^e (figure 20).

When $\varepsilon=0$, the two components of $-Y_r$ are positive and equal. Let us assume $T \gg T_0$ such that the components of $-(Y_r+S_r)$ are also positive. Thus the only possibility for the vector $-(Y_r+S_r)$ to be normal to the triangle C_r^e is to be normal on the side AB_e. Because the point $\beta_r^0=(\frac{1}{2},\frac{1}{2})$ does not belong to C_r^e, the stress $\sigma=K\varepsilon+\tau_{11}(T)(\beta_2-\beta_1)$ cannot be equal to 0. Thus the point ($\varepsilon=0$, $\sigma=0$) cannot be an equilibrium state at low temperature.

The possible equilibrium states at low temperature are shown in figure 21. Depending on the values of T and c there are different possibilities for the equilibrium domain at low temperature (see figures 20, 21 and figures 22, 23). On figure 23, the deformation ε_4 is the smallest deformation for which there exists S_r for which $-(Y_r+S_r)$ is normal to AB_e and OB_e (figure 22). The deformation ε_5 is also the largest deformation for which $-(Y_r+S_r)$ can be normal to C_r^e on AB_e and OB_e(figure 22).

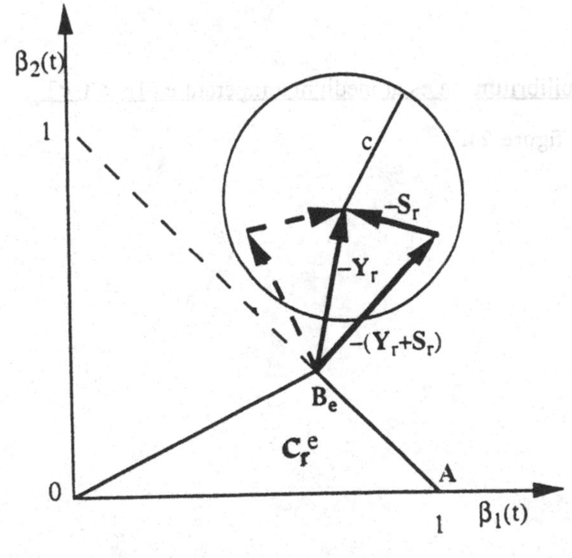

Figure 22
The vector - (Y_r+S_r) is normal to the triangle C_r^e. It can be normal to the sides $\mathbf{B_eA}$ and $\mathbf{OB_e}$ for different vectors $S_r\in S$.

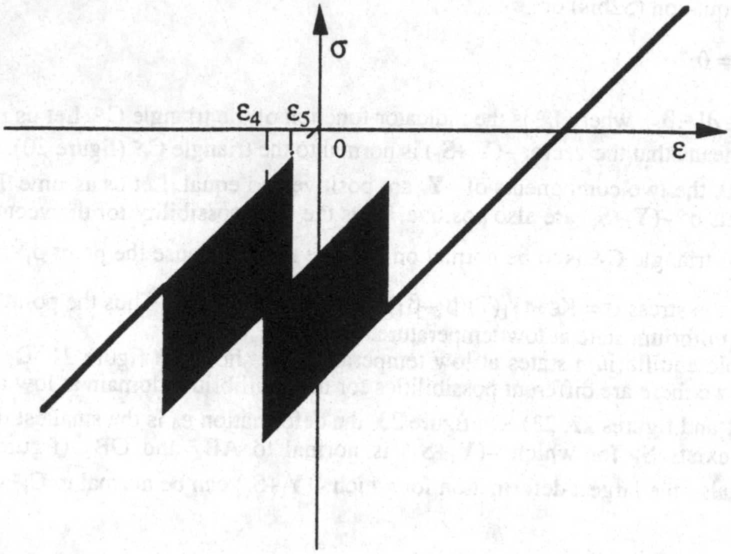

Figure 23
Equilibrium positions of an educated shape memory alloy at low temperature.

9.14.2. Equilibrium states at medium temperature ($T_0 < T < T_c$).

They are shown in figure 24.

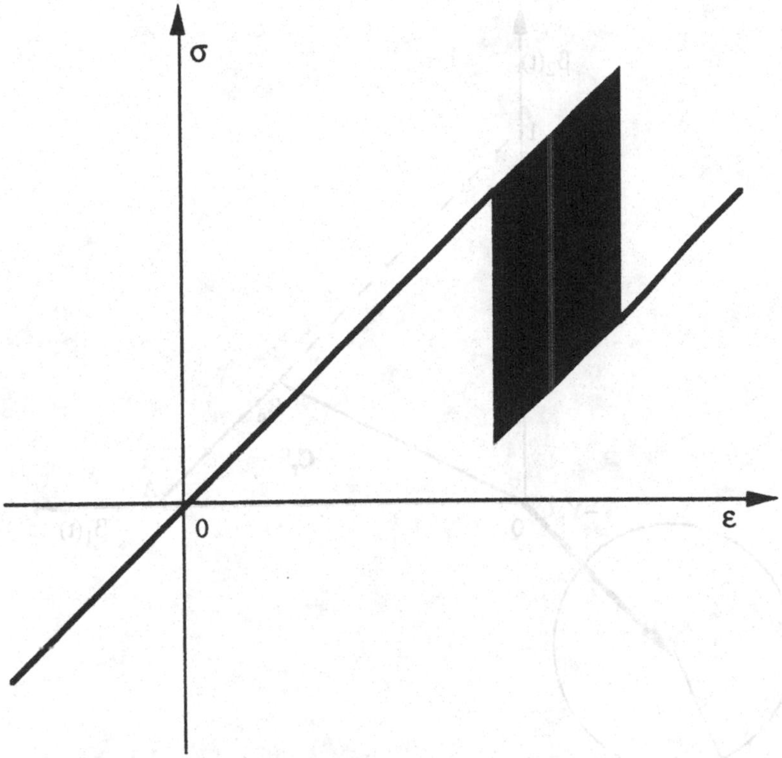

Figure 24
Equilibrium staes of an educated shape memory alloy at medium temperature.

9.14.3. <u>Equilibrium states at high temperature ($T > T_c$).</u>

The components of $-Y_r$ are equal and negative. We assume $T \gg T_c$ such that the components of $-(Y_r+S_r)$ are also negative (figure 25).

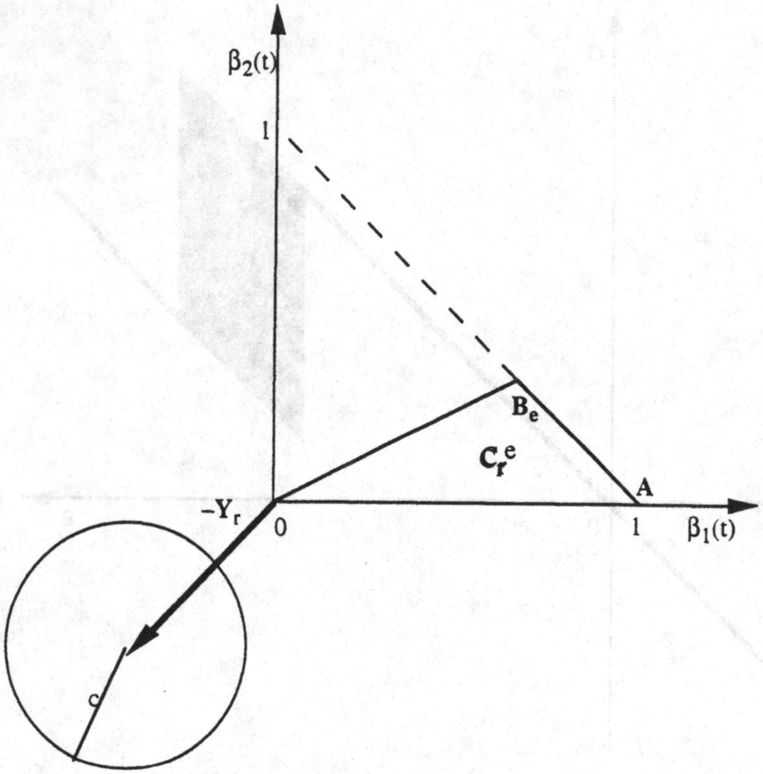

Figure 25
The vector – (Y_r+S_r) is normal to C_r^e at high temperature.

Thus the only possibility for this vector to be normal to C_r^e is to be normal at the point 0 ($\beta_r = 0$), (figure 25). The equilibrium states at high temperature satisfy (figure 26)

$$\sigma = K\varepsilon.$$

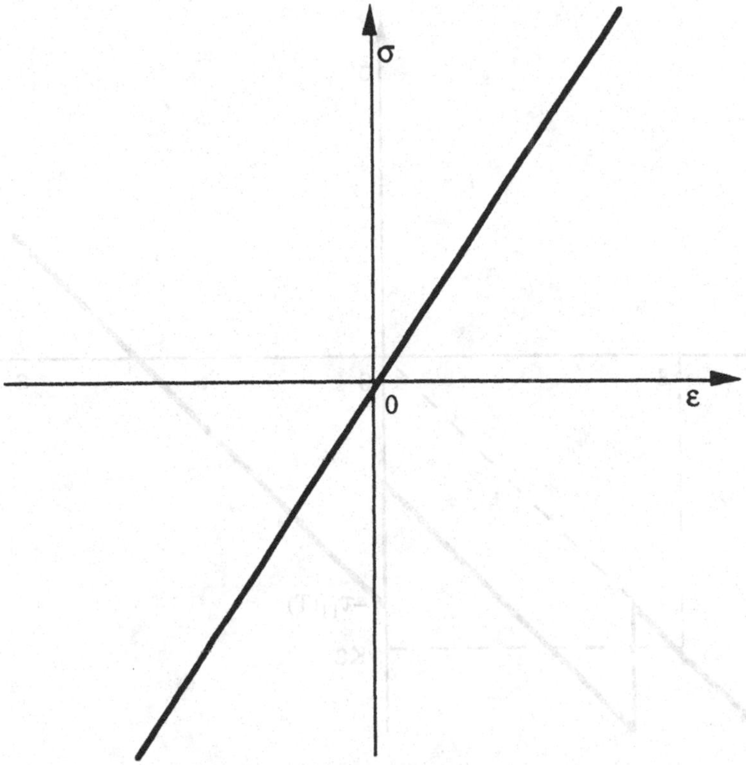

Figure 26
The constitutive law at high temperature.

9.14.4. Constitutive laws of a non-dissipative educated shape memory alloy

They are obtained by letting $c = 0$ in the previous figures showing the equilibrium positions of the dissipative shape memory. The consitutive laws for the low and medium temperature are shown in the following figures 27 and 28. The constitutive law at high temperature is the one shown in the previous figure 26.

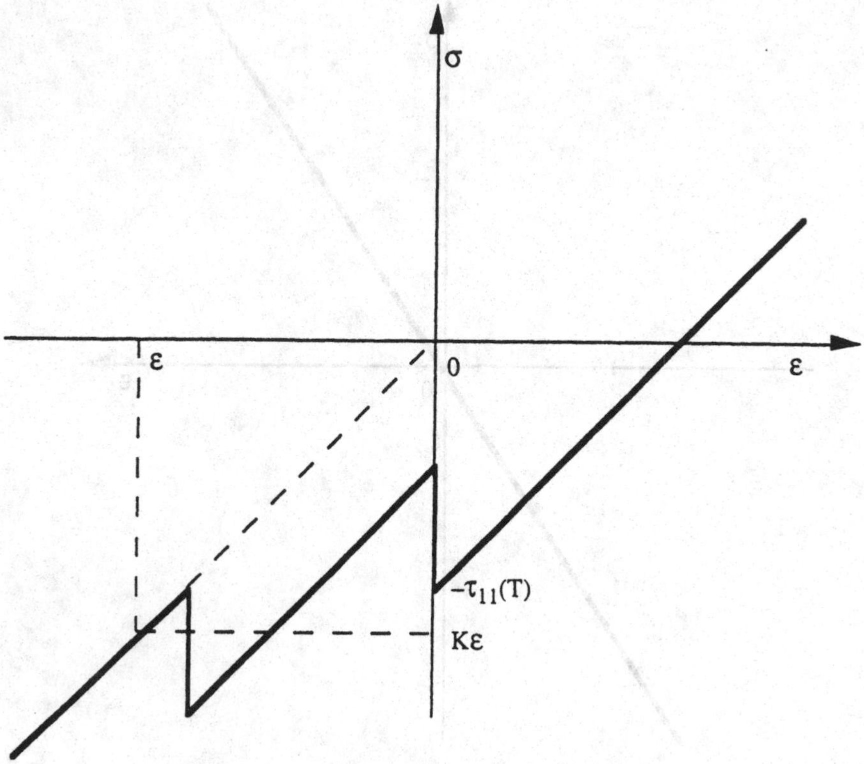

Figure 27
Constitutive law of a non-dissipative educated shape memory alloy at low temperature.

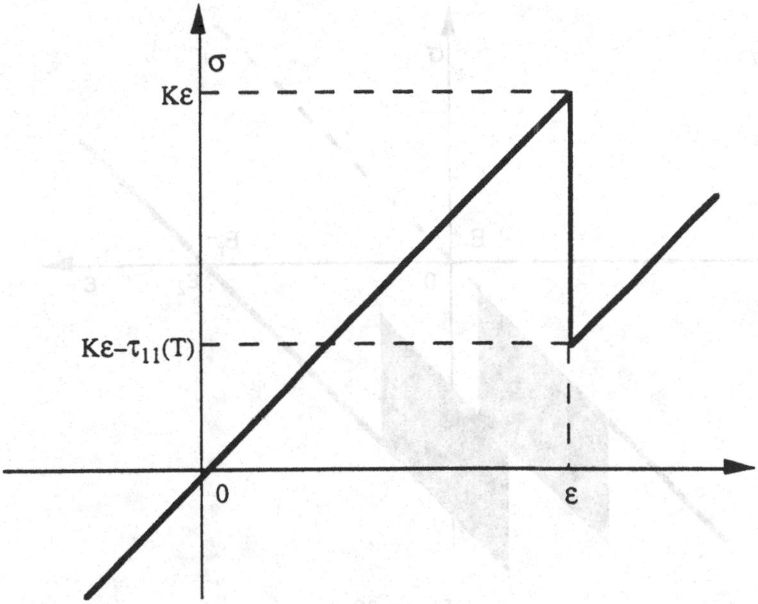

Figure 28
Constitutive law of a non-dissipative educated shape memory alloy at medium temperature.

9.14.5. The two shape memory effect .

Let us consider an unloaded ($\sigma = 0$) educated shape memory alloy at high temperature, T^+. We have $\varepsilon = 0$. Let us cool it, the only unloaded ($\sigma = 0$) equilibrium state at low remperature, T^-, is ($\sigma = 0, \varepsilon = \varepsilon_2$). If the temperature is again increased the alloys goes back to the state ($\sigma = 0, \varepsilon = 0$) and so on (figure 29). The alloys remembers two shapes: the state $E_r^+(\varepsilon = 0, \sigma = 0, \beta_3 = 1, T^+)$ and the state $E_r^-(\varepsilon = \varepsilon_2, \sigma = 0, \beta_1 = 1, T^-)$! One at high temperature and an other one at low temperature. The heat ΔQ which is received from the exterior during the phase change, for instance when going from E^- fo E^+, is

$$\Delta Q = \Delta e_r = e_r^+ - e_r^- = C\Delta T - \frac{1}{2}K(\varepsilon_2)^2 - T_c\tau_{11}\varepsilon_2 + l_a, \; (K = K_{1111})$$

with $\Delta T = T^+ - T^-$ and $\varepsilon_2 = \dfrac{\tau_{11}(T)}{K}$,

$$= C\Delta T + l_a - \{(T^-)^2 - T_c^2\}\frac{\tau_{11}^2}{2K}.$$

As we expect, it is positive because $T^- \leq T_c$.

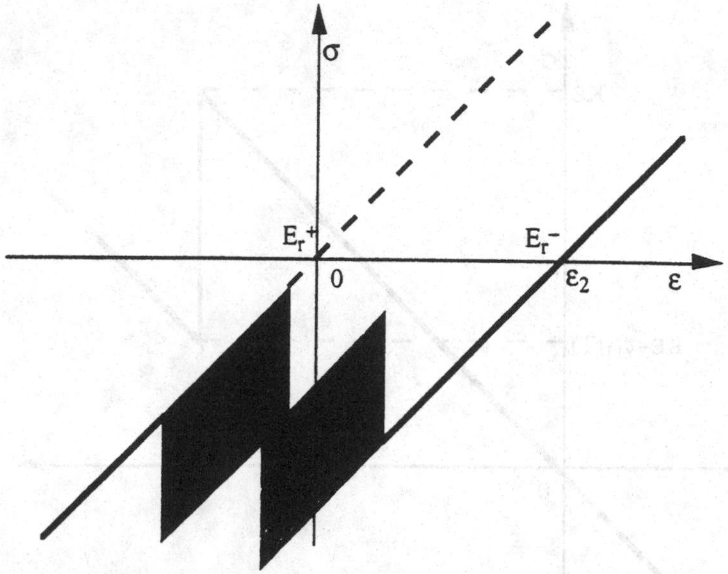

Figure 29

The two shape memory effect. The state goes from $E_r^+(T^+, \varepsilon = 0, \sigma = 0, \beta_3 = 1)$ to $E_r^-(T^-, \varepsilon = \varepsilon_2, \sigma = 0, \beta_1 = 1)$ and from E_r^- to E_r^+.

9.15. Smooth constitutive laws.

The equilibrium curves showing the stresses versus the deformations are non-smooth. They are angular. Experiments have shown that the pure martensite variants or mixtures, corresponding to the extreme points **A** and **B** in the triangle C_r. There is always a slight fraction of the second martensite at the point **A** (β_2 is very small but not zero). To take this property into account we replace the set of the possible mixture by a convex curvilinear triangle C_{rc} as shown in figure 30.

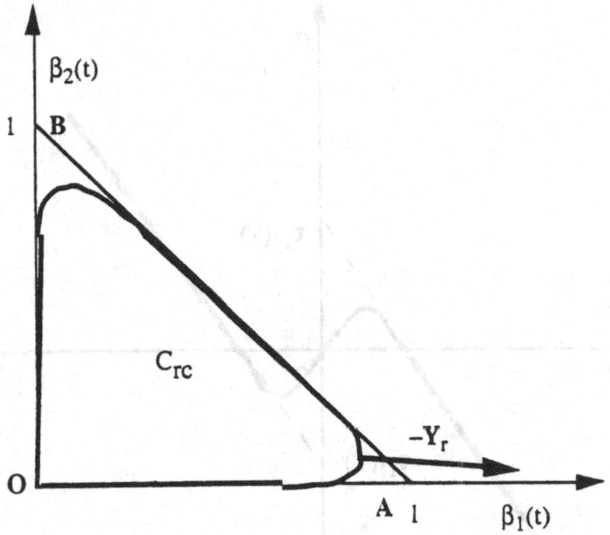

Figure 30
The curvilinear set of the possible mixtures of the martensites.

The effect of replacing the triangle C_r by the curvilinear C_{rc} on the extended free energy is to replace the indicator function I_r of C_r by the indicator function I_{rc} of C_{rc}. The effect on the constitutive laws is to replace the subdifferential ∂I_r by ∂I_{rc}. For instance the relations (40) is replaced by

(40bis), $\mathbf{B}_r^{ndr} \in \partial I_{rc}(\beta_r)$,

which means that the reaction $\mathbf{B}_r^{ndr}$ is normal to the convex curvilinear triangle C_{rc}. The equation of movement (46) where we assume no dissipation,

(46), $\mathbf{B}_r^{nd}(E_r)+\mathbf{B}_r^{ndr}(E_r,x,t) = \mathbf{B}_r = \mathbf{A}_r = 0$,

with

$$\mathbf{B}_r^{nd}(E_r) = \frac{\partial \Psi_r}{\partial \beta_r}(E_r) = \mathbf{Y}_r(E_r) = \begin{vmatrix} \Psi_{1r}(E_r)-\Psi_{3r}(E_r) \\ \Psi_{2r}(E_r)-\Psi_{3r}(E_r) \end{vmatrix},$$

gives

$-\mathbf{Y}_r \in \partial I_{rc}(\beta_r)$,

which means that the vector $-\mathbf{Y}_r$ is normal to the convex curvilinear triangle C_{rc} (figure 30). The result is a smoothing effect as shown in the figure 31 on the constitutive law at low temperature.

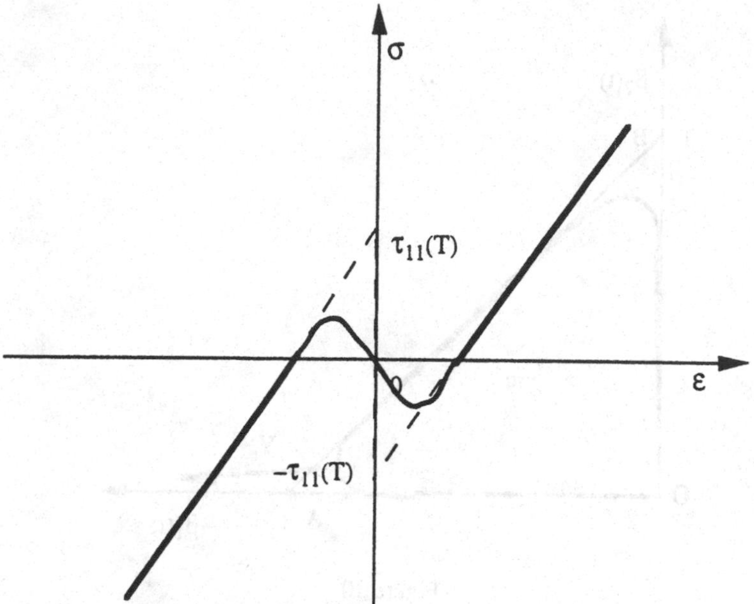

figure 31
The smooth constitutive law at low temperature.

10. Evolutions of structures made of shape memory alloys.

The evolution of a structure made of shape memory alloys, i.e. the computation of $E_r(\mathbf{x},t) = (\varepsilon(\mathbf{x},t),\beta_1(\mathbf{x},t),\beta_2(\mathbf{x},t),T(\mathbf{x},t))$ depending on the point **x** of the domain occupied by the structure and on the time t can be performed by solving numerically the set of partial differential equations resulting from the movement equations (2), (3), (5ter), (6ter), (or (5bis)) and the energy balance (7ter), (or (7bis), (8), the constitutive laws (40), (41), (42), (43), (44), (45) and (47). completed by convenient initial and boundary conditions. Let us mention the numerical results given by Wörshing [33] where numerous practical and theoretical results can be found. The coherence in term of mathematics of the set of partial differential equations has been proved in theoretical mathematical papers[8], [30], [31].

11. Conclusion.

The models we have described are able to account for the different features of the shape memory alloys macroscopic mechanical and thermal properties. We have used schematic free energies and pseudo-potentials of dissipation. There are many possibilities to sophisticate the basic choices we have made to take into account the practical properties of the shape memory alloys. Let us for instance mention that the pseudo-potential of dissipation can be modified to describe more precisely the hysteretical properties of the shape meùory alloys. There is no difficulty to have more than two martensites, for instance to take care of the 24 possible martensites: 22 more β's are to be introduced and the triangle C_r has to be replaced by the convex set $C_{24} = \{\beta \in \mathbf{R}^{24};\ 0 \leq \beta_i \leq 1$ for i =

1 to 24}!

Let us remark that the physical quantities to characterise an educated shape memory alloys are: K, C, l_a, $T_{\dot{a}}$, T_c, Σ, the two coordinates of $\mathbf{B}_e$ for the tree energy and c, k for the pseudo-potential of dissipation. It is not so many to have a complete multidimentional model which can be used for engineering purposes.

Let us also note the very important role of the internal constraints and of the reaction $\mathbf{B}^{ndr}$ to these internal constraints which are responsible for many properties. The way we have taken into account the internal constraints is general and can be developped in other circumstances [12], [13],[16], [32].

12. Appendix.

12.1. Convex function.

A convex function from **R** into $\bar{\mathbf{R}} = \mathbf{R}\cup\{+\infty\}$; (figure 32), [20] is an application f whose value can be $+\infty$ at some point (the value $-\infty$ is forbidden) such that

$$\forall x,y \text{ and } \forall\theta\in\,]0,1[,\qquad f(\theta x+(1-\theta)y) \le \theta f(x)+(1-\theta)f(y).$$

A convex function f from $\mathbf{R}^2$ into $\bar{\mathbf{R}} = \mathbf{R}\cup\{+\infty\}$is a function f such that

$$\forall\gamma\in\mathbf{R}^2,\ \forall\beta\in\mathbf{R}^2,\ \forall\theta\in\,]0,1[,\ f(\theta\gamma+(1-\theta)\beta) \le \theta f(\gamma)+(1-\theta)f(\beta).$$

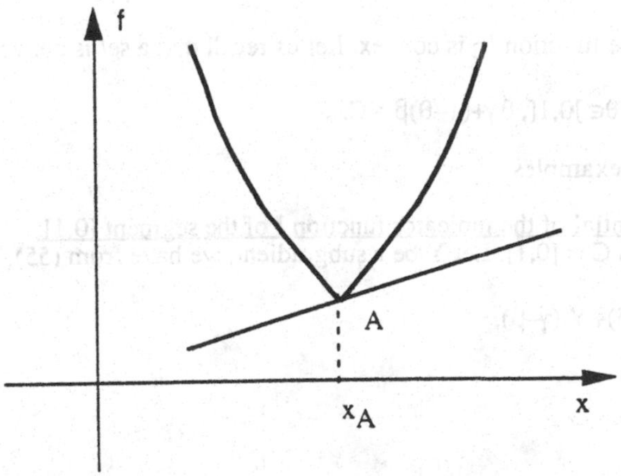

Figure 32
The convex function is not differentiable at point A. It has a generalized derivative: the slope of any line which passes through point A and is under the function f.

12.2. Subgradient and subdifferential of a convex function.

A convex function does not always have a derivative (for instance at the point A of figure 32) but it can have generalized derivatives, subgradients, which are the slopes of the lines which pass through point A and are always under the function. The set of all the sub-gradients is the subdifferential denoted $\partial f(x_A)$. The subgradients have the following property

(55), $\forall p \in \partial f(x_A), \forall y, f(y) \geq f(x_A)+(y-x_A)p.$

Let us remark that the set $\partial f(x)$ is empty if $f(x) = +\infty$ (because f(y) which is finite for some y cannot be greater than $+\infty$). Thus the relation $p \in \partial f(x)$ has two meanings: first that f(x) is finite and second that relation (55) is satisfied.

A vector $\mathbf{Y} \in \mathbf{R}^2$ is a subgradient of the convex function f from $\mathbf{R}^2$ into $\mathbf{R} = \mathbf{R} \cup \{+\infty\}$ at the point β if

(56), $\forall \gamma \in \mathbf{R}^2, f(\gamma) \geq f(\beta)+\mathbf{Y}.(\gamma-\beta).$

The set of all the subgradients of f at the point β is the subdifferential, $\partial f(\beta)$.

12.3. <u>Indicator function of a set.</u>

The indicator function I_C of a set C is defined by

$I_C(\gamma) = 0$, if $\gamma \in C$,

$I_C(\gamma) = +\infty$, if $\gamma \notin C$.

If the set C is convex, the function I_C is convex. Let us recall that a set is convex iff

$\forall \gamma \in C, \forall \beta \in C, \forall \theta \in]0,1[, \theta\gamma+(1-\theta)\beta \in C.$

Let us give some useful examples.

12.4. <u>Subdifferential of the indicator function I of the segment [0,1].</u>
The convex set is C = [0,1]. Let Y be a subgradient, we have from (55), (figure 33),

$\forall \gamma \in [0,1], 0 \geq I(\beta)+Y.(\gamma-\beta).$

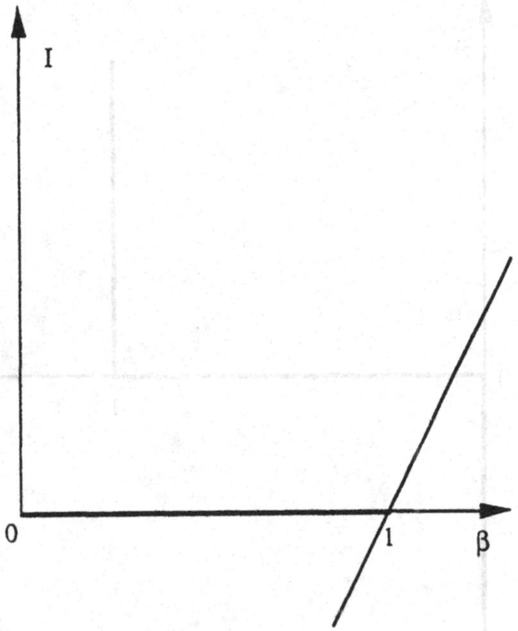

Figure 33
A subgradient of the indicator function I of the segment [0,1] at the point $\beta = 1$ is the slope of a line passing through the point (1,0) and which is under the function I.

It results (figure 34),

$$\partial I(\beta) = \emptyset, \text{ if } \beta \notin [0,1],$$

and

$$\partial I(\beta) = \{0\}, \text{ if } \beta \in]0,1[,$$

$$\partial I(1) = \mathbf{R}^+,$$

$$\partial I(0) = \mathbf{R}^-.$$

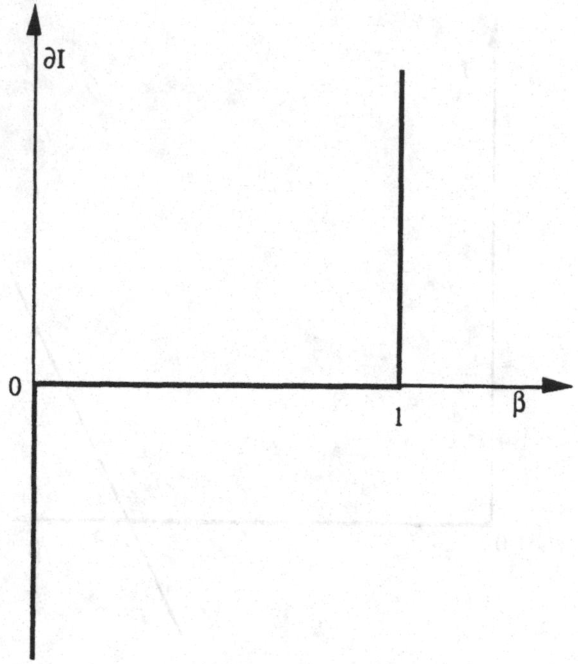

Figure 34
The graph $\partial I(\beta)$.

12.5. Subdifferential of the indicator function I_{0r} of the origin of **R**.
The convex set is $C = \{0\}$. Let Y be a subgradient, we have from (55),

$$0 \geq I_{or}(x)+Yx.$$

It results that

$$\partial I_{or}(x) = \varnothing, \text{ if } x \neq 0,$$

$$\partial I_{or}(0) = \mathbf{R};$$

because any subgradient at the origin x = 0 is such that $0 \geq Y0$, (figure 35).

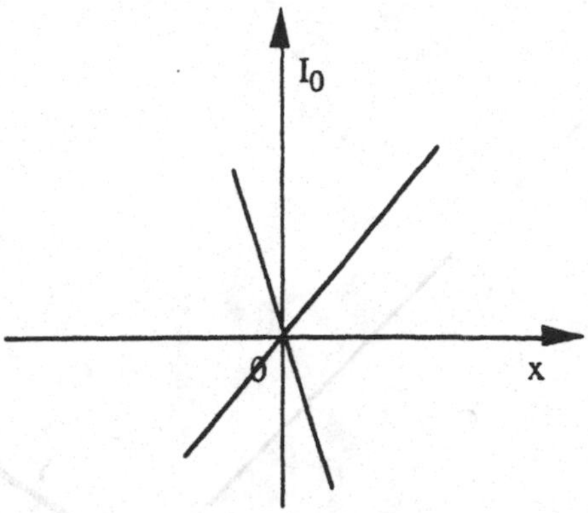

Figure 35
Any subgradient of the indicator function of the origin is the slope of a linear function.

12.6. Subdifferential of the indicator function I_r of a triangle C.

Let **Y** be a subgradient, we have from (56),

$\forall \gamma \in C,\ 0 \geq I_r(\beta)+\mathbf{Y}.(\gamma-\beta).$

It results that

$\partial I_r(\beta) = \varnothing$, if $\beta \notin C$,

and that the vector **Y** is normal to the triangle C at the poinr β if $\beta \in C$ (figure 36).

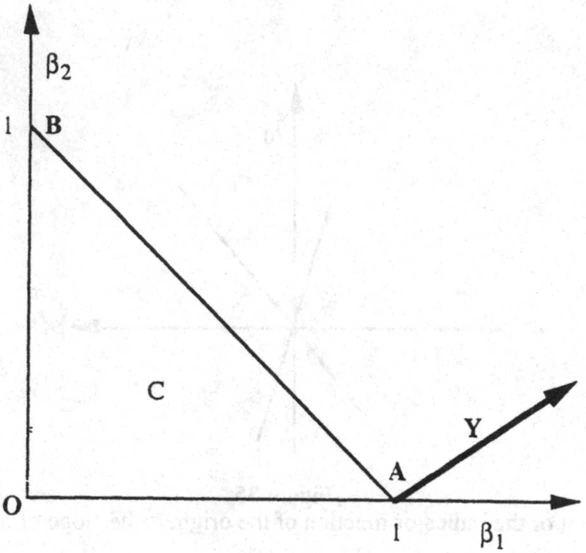

Figure 36
The triangle C with vertices **O** = (0,0), **A** = (1,0) and **B** = (0,1). The vector **Y** is normal to the triangle C at the point **A**.

Thus we have

$\partial I_C(\beta) = \{0,0\}$, if β is in the interior of the triangle,

$\partial I_C(\beta) = \{(Y,Y)\ ;\ Y \geq 0\}$, if β is on the side AB ;

$\partial I_C(\beta) = \{(0,Y)\ ;\ Y \leq 0\}$, if β is on the side OA ;

$\partial I_C(\beta) = \{(Y,0)\ ;\ Y \leq 0\}$, if β is on the side OB ;

$\partial I_C(\mathbf{O}) = \{(Y_1,Y_2)\ ;\ Y_1 \leq 0,\ Y_2 \leq 0\}$;

$\partial I_C(\mathbf{A}) = \{(Y_1,V_2)\ ;\ Y_1 \geq 0,\ Y_2 \leq Y_1\}$;

$\partial I_C(\mathbf{B}) = \{(Y_1,Y_2)\ ;\ Y_2 \geq 0,\ Y_1 \leq Y_2\}$.

13. <u>References.</u>

[1] R.Abeyaratne, J.Knowles, A continuum model of thermoelastic solid capable of undergoing phase transitions, J. Mech. Phys. Solids, vol 41, n°3, pp. 541-571, 1993.

[2] M.Achenbach, A model for an alloy with shape memory, International Journal of plasticity, vol.5, pp. 371-395, 1989.

[3] M.Achenbach, I.Müller, Simulation of material behavior of alloys with shape memory, Arch. Mech., 37, 6, pp. 573-585, 1985.

[4] C.Berriet, C.Lexcellent, B.Raniecki, A.Chrysochoos, Pseudoelastic behaviour analysis by infrared thermography and resistivity measurements of polycristalline shape memory alloys, ICOMAT 92, Monterey, 1992.

[5] A.Chrysochoos, H.Pham, O.Maisonneuve, Une analyse expérimentale du comportement d'un alliage à mémoire de forme de type Cu-Zn-Al, C. R. Acad. Sci., tome 316, série II, pp.1031-1036, Paris, 1993.

[6] A.Chrysochoos, M.Löbel, O.Maisonneuve, Couplages thermécaniques du comportement pseudoélastique d'alliages Cu-Zn-Al et Ni-Ti, C. R. Acad. Sci., tome 320, série IIb, pp. 217-223, Paris, 1994.

[7] J.M.Ball, R.D.James, Theory for the microstructure of martensite and applications, Proc. Inter. Conf. on Martensitic Transformations, C.M.Wayman, J.Perkings eds, Monterey, 1992.

[8] P.Colli, M.Frémond, A.Visintin, Thermomechanical evolution of shape memory alloys, Quaterly of applied mathematics, Vol XLVIII, n°1, pp. 31-47, 1990.

[9] F.Falk, Landau theory and martenstic phase transition, Journal de Physique; Colloque C4, Supplément au n°12, tome 43, 1982.

[10] M.Frémond, Matériaux à mémoire de forme, C. R. Acad. Sci., tome 304, série II, n°7, pp. 239-244, Paris, 1987.

[11] M.Frémond, Shape memory alloys. A thermomechanical model, in Free boundary problems: theory and application (K.H.Hoffmann, J.Spreckels eds), Pittman, Longman, Harlow, 1988.

[12] M.Frémond, Sur l'inégalité de Clausius-Duhem, C. R. Acad. Sci., tome 311, serie II, pp. 757-762, Paris, 1990.

[13] M.Frémond, B.Nedjar, Endommagement et principe des puissances virtuelles, C. R. Acad. Sci., tome 317, serie II, n° 7, pp. 857-864, Paris, 1993.

[14] P.Germain, Mécanique, Ellipses, Paris, 1986.

[15] G.Guénin, Alliages à mémoire de forme, Techniques de l'ingénieur, M 530, Paris, 1986.

[16] H.Ghidouche, N.Point, Unilateral contact with adherence, in Free boundary problems: theory and application (K.H.Hoffmann, J.Spreckels eds), Pittman, Longman, Harlow, 1988.

[17] R.D.James, D.Kinderleherer, Theory of diffusionless phase transformation, Lectures notes in physics 344 (M.Rascle, D.Serre, M.Slemrod ed.s), Springer-Verlag, Heidelberg, 1990.

[18] S.Leclercq, De la modélisation thermomécanique et de l'utilisation des alliages à mémoire de forme, thèse de l'Université de Franche-Comté, Besançon, 1995.

[19] C.Lexcellent, C.Licht, Some remarks on the modelling of the thermomechanical behaviour of shape memory alloys, Journal de Physique, Colloque C4, vol. 1, pp. 35-39, 1991.

[20] J.J.Moreau, Fonctionnelles convexes. Séminaire sur les équations aux dérivées partielles, Collège de France, Paris, 1966.

[21] J.J.Moreau, Sur les lois de frottement, de viscosité et de plasticité, C. R. Acad. Sci., vol. 271, pp. 608-611, Paris, 1970.

[22] I.Müller, H.Xu, On pseudoelastic hysteresis, Acta Metall. Mater., 39, 1991.

[23] I.Müller, Pseudo elasticity in shape memory alloys. An extreme case of thermoelasticity. Proc. Thermoelasticita Finita, Acc. Naz. dei Lincei, May/June 1985.

[24] M.Niezgodka, J.Sprekels, Convergent numerical approximation of the thermomechanical phase transitions in shape memory alloys, to appear.

[25] Nguyen Quoc Son, Z. Moumni, Sur une modélisation du changement de phases solides, C. R. Acad. Sci., serie II, Paris, 1995.

[26] E.Patoor, M.Berveiller, Les alliages à mémoire de forme, Hermès, Paris, 1990.

[27] H.Pham, Analyse thermomécanique d'un alliage à mémoire de forme de type Cu-Zn-Al, thèse de l'Université des Sciences et des Techniques du Languedoc, Montpellier, 1994.

[28] P.Podio-Guidugli, G.V.Caffarelli, Equilibrium phases and layered phase mixtures in elasticity, to appear.

[29] B.Raniecki, C.Lexcellent, K.Tanaka, Thermodynamics models of pseudoelastic behaviour of shape memory alloys, Arch. Mech., 44, 3, pp. 261-284, 1992.

[30] J.Sprekels, Global existence for thermomechanical process with non convex free energies of Ginzburg-Landau form, J. Math. Anal. Appl..

[31] J.Sprekels, Shape memory alloys: mathematical models for a class of first order solid-solid phase transition in metals, Control and cybernetics, vol. 19, n°3, 4, 1990.

[32] J.M.T. Tien, L'adhèrence des solides, thèse de l'Université Pierre et Marie Curie, Paris, 1990.

[33] G.Wörsching, Ph.D thesis, Augsburg University, 1994.

DEVELOPMENT AND CHARACTERIZATION OF SHAPE MEMORY ALLOYS

S. Miyazaki
University of Tsukuba, Tsukuba, Japan

Abstract:

Recent development of shape memory alloys is reviewed, emphasis being placed on the Ti-Ni, Cu-based and ferrous alloys which are considered as practical materials for applications among many shape memory alloys. Crystal structures of the parent and martensitic phases are described, and the crystallography of the martensitic transformations is also briefly explained. The origin of the shape memory effect and the shape memory mechanisms are discussed on the basis of the crystal structure and the crystallography of the martensitic transformations. Since an applied stress also induces the martensitic transformations, successive stages of the stress-induced martensitic transformations are reviewed briefly in Cu-based and Ti-Ni alloys, which show martensite-to-martensite transformations upon loading. Then, the transformation and mechanical characteristics of the shape memory alloys are reviewed in detail; i.e. phase diagrams, transformation temperatures, transformation process, stres-induced transformation, aging effects, cycling effects, fracture, fatigue, grain refinement, two-way shape memory effect, and so on. Recent develpment of sputter-deposited Ti-Ni thin films is also introduced.

Key words:
Shape memory effect, Pseudoelasticity, Superelasticity, Martensite, Martensitic transformation, R-phase, Rhombohedral phase, Ti-Ni, Ni-Ti, Cu-Al-Ni, Cu-Zn-Al, Shape memory alloy, Ferrous shape memory alloy, Fatigue, Crack propagation, Fracture, Single crystal, Bi-crystal, Two-way shape memory effect, Thin film

1. Introduction

The shape memory effect (SME) appears in some special alloys which show crystallographically reversible martensitic transformations. The martensitic transformation is accompanied by a large shear-like deformation associated with a diffusionless structural change; the deformation generally amounts to about 20 times more than the elastic deformation. The martensite is deformable and it can also be induced from the parent phase by loading, both deformation modes being associated with no permanent strain in the shape memory alloys (SMAs). Thus, a large deformation induced in the SMAs can recover perfectly by heating to temperatures above the reverse-transformation finish temperature (*Af*) after unloading (shape memory effect or SME) or simply by unloading at temperatures above Af (pseudoelasticity (PE) or superelasticity (SE)).

The SME was first found in a Au-Cd alloy in 1951 [1], and then in In-Tl alloy in 1953 [2,3]. However, the possibility for using the SME in actual applications was realized after the SME was found in a Ti-Ni alloy in 1963 [4]. Since the Ti-Ni alloy has many complicated features and difficulty in making single crystals, the basic understanding of this alloy was not possible until thc carly 1980s. On the other hand, a Cu-Al-Ni alloy was also found to reveal the SME in 1964 [5] and in this alloy it was demonstrated that the SME is closely related to the thermoelastic martensitic transformation [6]. Since then, the basic understanding of the origin of the SME[6], the shape memory mechanism(7-9), the crystallography of the stress-induced martensitic transformation [10,11] has been established for Cu-based alloys in the 1970s. This is partly due to an easiness of making single crystals of these alloys.

However, the Ti-Ni alloy has been the most important material for applications, because the Cu-based alloys are brittle in a polycrystalline state [12]. In the 1980s, the basic understanding of the thermomechanical treatment [13,14], the deformation behavior[15,20], the shape memory mechanism [21,22], the nature of the R-phase[23,31], the crystallography[32,36] of both the R-phase and the martensitic transformations, and the phase diagram including the crystal structure of metastable precipitates[37-40] has been obtained for the Ti-Ni alloys.

In the same 1980s, many patents of applications using the SME and PE have been applied for; now the total number of the applied patents amounts to more than 15,000[41]. At this stage, the price of the shape memory alloys is also one of the key factors for applications. Recently, it was also found that some ferrous alloys also exhibit perfect SME under certain conditions[42-45].

This chapter will first review the fundamental aspects of SMAs, i.e. the crystal structures, the crystallography of martensitic transformations, origins of SME, shape memory mechanism, successive stress-induced transformation and so on. And then most part of this paper will be devoted to reviewing the recent development and characterization of the Ti-Ni alloys, Cu-based alloys and ferrous alloys.

There are many good text books[46-52] and reviews [8,53-62] on SME and PE, and they will be useful for obtaining a deeper understanding of the the fundamentals of the SMAs.

2. Fundamental Aspects of Shape Memory Alloys

2.1 Crystal Structure

Most SMAs have superlattice structures, the sublattices of the parent phases being body-centered cubic (BCC), as shown in Table 1. The parent phases of some other alloys (In-Tl, In-Cd, Mn-Cu) have disordered lattices, i.e. face-centered cubic (FCC). If we treat these alloys as exceptional cases, the parent phases of all alloy listed in the table have superlattices associated with the BCC structure, and these are classified as β-phase alloys, when electron per atom ratio (e/a) is close to 1.5.

Table 1 Non-ferrous alloys exhibiting perfect shape memory effect and pseudoelasticity

Alloy	Composition (at%)	Structure change	Temperature hysteresis (K)	Ordering
Ag-Cd	44~49Cd	B2—2H	~15	Ordered
Au-Cd	46.5~50Cd	B2—2H	~15	Ordered
Cu-Zn	38.5~41.5Zn	B2—9R, rhombohedral M9R	~10	Ordered
Cu-Zn-X (X=Si, Sn, Al, Ga)	A few at%	B2 (DO_3)—9R, M9R (18R, M18R)	~10	Ordered
Cu-Al-Ni	28~29Al 3~4.5Ni	DO_3—2H	~35	Ordered
Cu-Sn	~15Sn	DO_3—2H, 18R	—	Ordered
Cu-Au-Zu	23~28Au 45~47Zn	Heusler—18R	~6	Ordered
Ni-Al	36~38Al	B2—3R	~10	Ordered
Ti-Ni	49~51Ni	B2—monoclinic B2—rhombohedral	20~100 1~2	Ordered
In-Tl	18~23Tl	FCC—FCT	~4	Disordered
In-Cd	4~5Cd	FCC—FCT	~3	Disordered
Mn-Cu	5~35Cu	FCC—FCT	—	Disordered

Table 2 Ferrous alloys exhibiting perfect or nearly perfect shape memory effect

Alloy	Composition	Structure change	Temerature hystresis (K)	Ordering
Fe-Pt	~25 at% Pt	$L1_2$—ordered BCT	Small	Ordered
Fe-Pd	~30 at% Pd	FCC—FCT	Small	Disordered
Fe-Ni-Co-Ti	33 % Ni, 10 % Co, 4 % Ti (wt%)	FCC—BCT	Small	Disordered
Fe-Ni-C	31 % Ni, 0.4 % C (wt%)	FCC—BCT	Large	Disordered
Fe-Mn-Si	~30 % Mn, ~5 % Si (wt%)	FCC—HCP	Large	Disordered
Fe-Cr-Ni-Mn-Si-Co	~10 % Cr, <10 % Ni, <15% Mn, <7 % Si, <15 % Co (wt%)	FCC—HCP	Large	Disordered

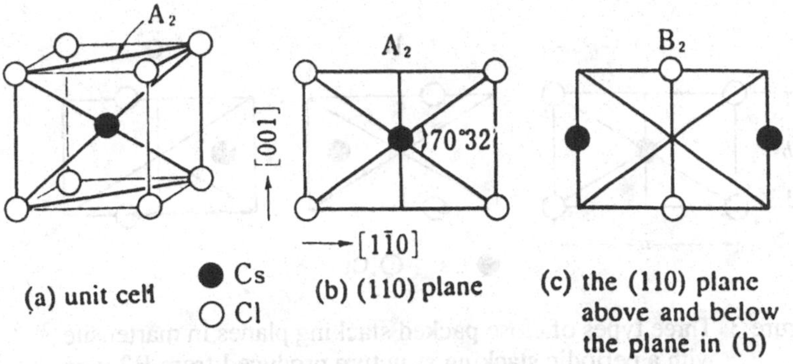

Figure 1 Crystal structure of B2 type structure (β_2 parent phase). A_2,B_2: (110) planes alternately stacked[46].

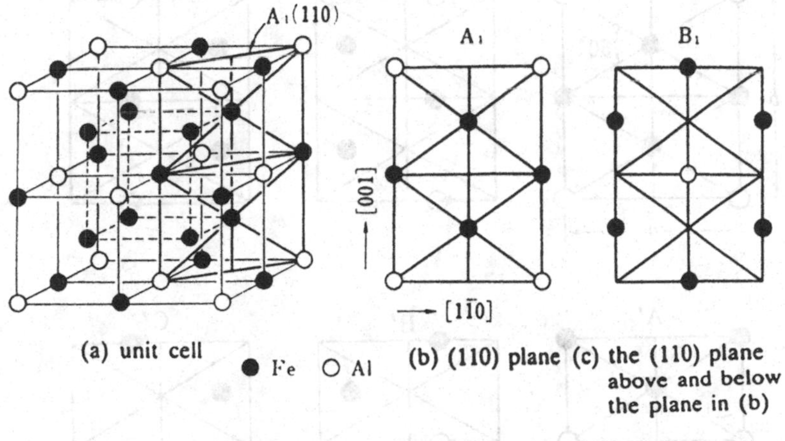

Figure 2 Crystal structure of DO3 type structure (β_1 parent phase). A_1,B_1: (110) planes alternately stacked[46].

Recently, it has been found that some ferrous alloys also show the SME as shown in Table 2. If we exclude Fe_3Pt, which has a superlattice associated with the FCC structure, all ferrous alloys have disordered structures (FCC) in the parent phases.

Most of the β-phase alloys are divided into two types according to the superlattice or composition ratio. One type is denoted by β_2-phase, which has a CsCl-type B2 superlattice and about 50:50 composition ratio. The other type is denoted by β_1-phase, which has a Fe_3Al-type DO_3 superlattice and about 75:25 composition ratio. Crystal

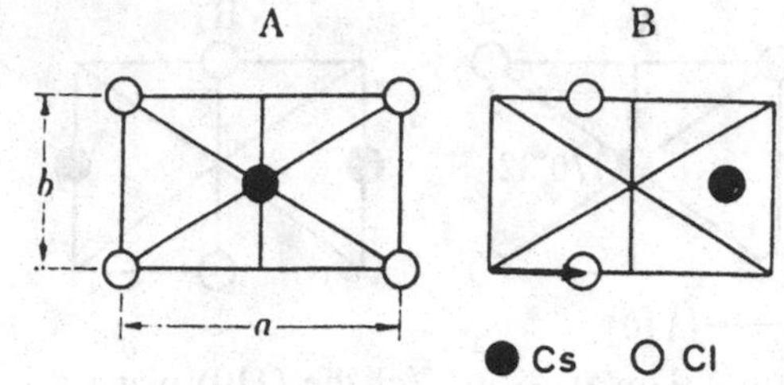

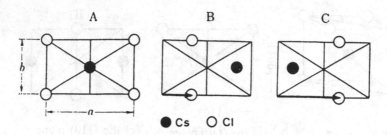

Figure 3 Three types of close packed stacking planes in martensite with a periodic stacking structure produced from B2 type β_2 parent phase [46].

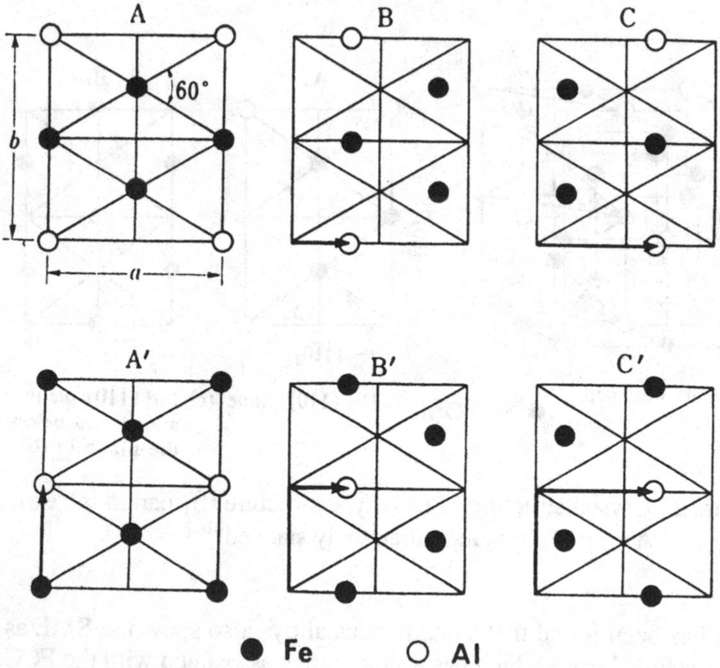

Figure 4 Six types of close packed stacking planes in martensite with a periodic stacking structure produced from a DO_3-type β_1 parent phase [46].

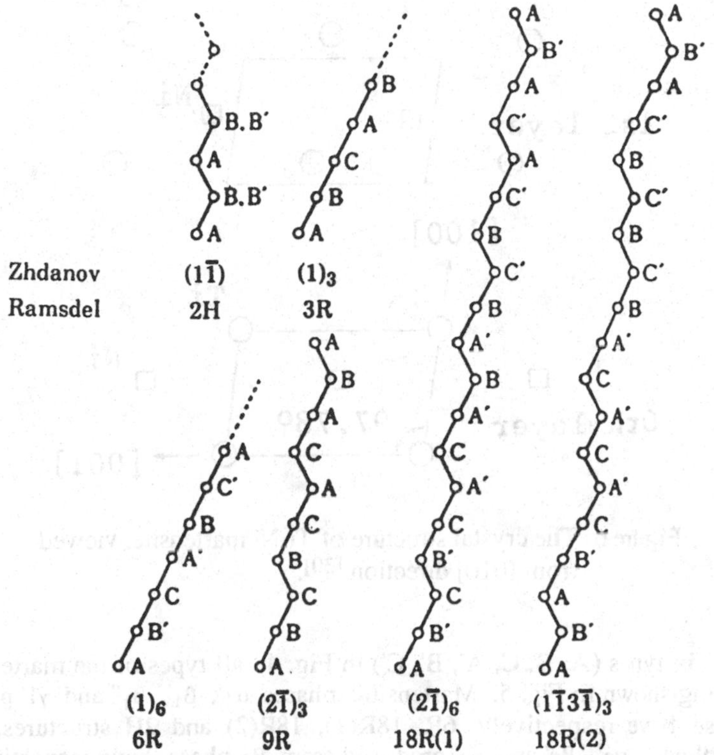

Figure 5 Periodic stacking structures with various stacking sequences [35].

structures for both cases are shown in Figs. 1 and 2, where (a) illustrates the three-dimensional structure, (b) the arrangement of the atoms in the (110) plane, and (c) the arrangement in the (110) plane above or below (b). The cubic structure in (a) can be constructed by alternate stacking of the planes in (b) and (c) [46].

The martensitic phase of the β-phase alloys also have superlattices which inherited from the crystal structure of the parent phase. Upon martensitic transformation, each (110) plane of the parent phase deforms nearly to a hexagonal network and shifts in the [1$\bar{1}$0] direction by a shear, resulting in the arrangement of the atoms in the (110) plane as shown in Fig. 3 or 4. Accordingly, the resulting crystal structures of the martensites can be constructed by stacking the three types atomic planes (A, B, C) in

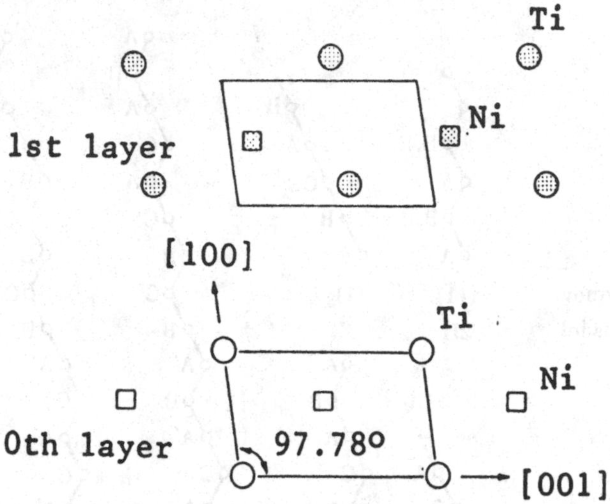

Figure 6 The crystal structure of Ti-Ni martensite, viewed from [010] direction [30].

Fig. 3 or the six types (A, B, C, A', B', C') in Fig. 4, all types of the martensite crystal structures being shown in Fig. 5. Martensitic phases α_1', β_1', β_1'' and $\gamma 1'$ produced from β_1-phase have respectively 6R, 18R(1), 18R(2) and 2H structures, while martensitic phases α_2', β_2' and γ_2' produced from β_2-phase have respectively 3R, 9R and 2H structures.

The martensitic phase of the Ti-Ni has a different structure, which is three-dimensionally close packed as shown in Fig. 6, although the parent phase has a CsCl-type B2 superlattice. The Ti-Ni alloy also shows another phase transition prior to the martensitic transformation according to heat-treatment. This phase transition consists of two processes; i.e., B2-incommensurate phase-commensurate phase (rhombohedral phase or R-phase) transition. The lattice does not change upon the former transition, while it changes to the rhombohedral lattice upon the latter transition. The R-phase can be formed by elongating along any one of the <111> directions of the B2 structure as shown in Fig. 7.

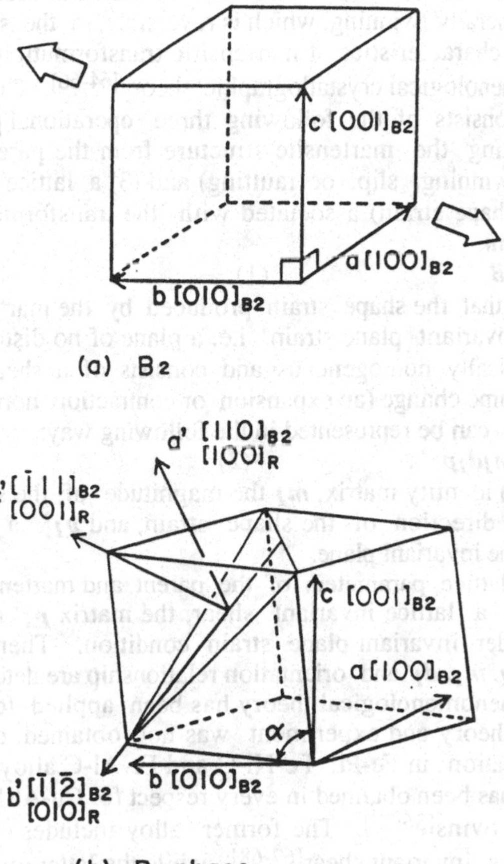

Figure 7 Unit cell of (a) the parent B2 phase and (b) the R-phase. The principal axes in the lattice deformation associated with the R-phase transition is also shown in (b) [31].

2.2 Crystallography of Martensitic Transformation

Martensitic transformation occurs in such a way that the interface between the martensite variant and parent phase becomes an undistorted and unrotated plane (invariant plane or habit plane) in order to minimize the strain energy. In order to form such a martensite variant (habit-plane variant), it is necessary to introduce a lattice invariant shear such as twins, dislocations or stacking faults[64-66]. The lattice invariant shear is generally twinning, which is reversible, in the shape memory alloys.

Crystallographic characteristics of martensitic transformations are now well understood by the phenomenological crystallographic theory[64-66]. This theory describes that the transformation consists of the following three operational processes: (1) a lattice deformation $\boldsymbol{B}$ creating the martensite structure from the parent phase, (2) a lattice invariant shear $\boldsymbol{P_2}$ (twinning, slip, or faulting) and (3) a lattice rotation $\boldsymbol{R}$. Thus, the total strain (or the shape strain) associated with the transformation is written in the following matrix form:

$$\boldsymbol{P_1} = \boldsymbol{RP_2B} \qquad (1)$$

This theory requires that the shape strain produced by the martensitic transformation is described by an invariant plane strain, i.e. a plane of no distortion and no rotation, which is macroscopically homogeneous and consists of a shear strain parallel to the habit plane and a volume change (an expansion or contraction normal to the habit plane). Thus, the shape strain can be represented in the following way:

$$\boldsymbol{P_1} = \boldsymbol{I} + \boldsymbol{m_1 d_1 p_1}' , \qquad (2)$$

where $\boldsymbol{I}$ is the (3x3) identity matrix, $\boldsymbol{m_1}$ the magnitude of the shape strain, $\boldsymbol{d_1}$ a unit column vector in the direction of the shape strain, and $\boldsymbol{p_1}'$ a unit row vector in the direction normal to the invariant plane.

If we know the lattice parameters of the parent and martensitic phases, a lattice correspondence and a lattice invariant shear, the matrix $\boldsymbol{p_1}'$ can be determined by solving Eq. (1) under invariant plane strain condition. Then, all crystallographic parameters such as $\boldsymbol{P_1}$, $\boldsymbol{m_1}$, $\boldsymbol{d_1}$ and orientation relationship are determined.

Although the phenomenological theory has been applied to many alloys, overall agreement between theory and experiment was not obtained, except for Au-Cd and {3,10,15} transformation in Fe-Pt, Fe-Ni-C and Fe-Al-C alloys. However, recently excellent agreement has been obtained in every respect for Cu-Al-Ni and Ti-Ni alloys by introducing Type II twins[36,37]. The former alloy includes the $<111>_M$ Type II twinning as the lattice invariant shear[67-68], while the latter one the $<011>_M$ Type II twinning[32,36].

2.3 Origins of Shape Memory Effect

The driving force for the reverse-transformation is the difference between the chemical free energy of the parent and martensitic phases above A_s, and the complete shape recovery lies in that the original orientation of the parent phase can be restored (crystallographic reversibility). The crystal structures of non-ferrous shape memory

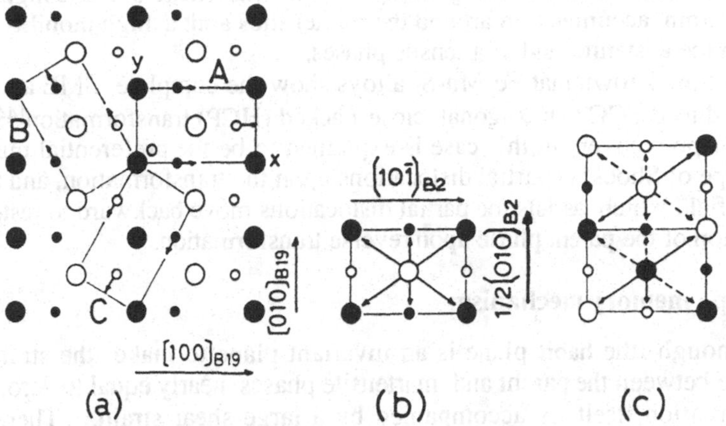

Figure 8(a) Three possible lattice correspondences in reverse transformation of the B2 to B19 transformation.
(b) Parent phase (B2 structure) resulting from reverse transformation with lattice correspondence A.
(c) Parent phase (different from B2 structure) resulting from reverse transformation with lattice correspondence B [69].

alloys exhibiting thermoelastic martensitic transformations are ordered, except for the alloys exhibiting FCC-FCT transformations.

The origin of complete reversibility upon reverse transformation is explained for an ordered alloy in Fig. 8 [69]. If the ordered arrangement of the atoms is ignored, there are three equivalent lattice correspondences; these are represented by the rectangles marked A, B and C in Fig. 8(a). If the reverse transformation takes the path A in the figure, the original B2 superlattice structure can be restored as shown in (b). However, if the reverse transformation takes the path B or C, the product phase will be as shown in (c), which is not the B2 structure anymore. In this case, the chemical free energy of the product phase is higher than that of the B2 structure, and thus the reverse transformation taking the path B or C becomes unfavorable. Therefore, the original orientation of the parent phase can be restored.

The FCC-FCT transformations in In-Tl and other alloys also occurs in a crystallographically reversible manner in spite of disordered structures. This is because the lattice deformation upon transformation is extremely small, and hence taking the path to the original orientation of the parent phase is easier than taking the other paths to the different orientations.

Most of ferrous shape memory alloys also have special situations for restoring the original orientation of the parent phase upon reverse transformation, although they have disordered structures. In the Fe-Ni-Co-Ti alloy, fine coherent γ_2 precipitates with $L1_2$ order are formed in the austenite matrix by ausaging[43]. Thus, the hardness of the austenite and the tetragonality of the martensite become high, resulting in a elastic strain accumulation around the martensites and a high mobility of the interface between the austenite and martensite phases.

It is now known that Fe-Mn-Si alloys show the complete SME associated with the stress-induced FCC to hexagonal close-packed (HCP) transformation[44,45]. The origin of the shape recovery in this case is explained to be the preferential multiplication of a single ype of Shockley partial dislocations upon the transformation, and they accumulate a stress field which assists the partial dislocations move backward to restore the original orientation of the parent phase upon reverse transformation.

2.4 Shape memory mechanism

Although the habit plane is an invariant plane to make the strain energy of the interface between the parent and martensite phases nearly equal to zero, the martensitic transformation itself is accompanied by a large shear strain. Therefore, the self-accommodation of martensite variants becomes also important in order to minimize the overall strain energy, resulting in nearly no macroscopic shape change in a specimen.

By applying a force, the most favorable variant outgrows in a self-accommodating morphology, and finally the maximum recoverable strain is attained; the strain remains even after unloading. By heating the deformed specimen to a temperature above *Af*, the original shape recovers completely by reversible reverse transformation.

In order to understand the shape memory mechanism, it is necessary to clarify the above processes; i.e. (a) self-accommodation, (b) variant coalescence to produce the most favorable variant upon loading, and (c) shape recovery upon heating. These processes will be explained in the following.

2.4.1 Self-accommodation

Three types of self-accommodating morphologies have been observed; i.e. a diamond-shaped morphology in β-phase alloys which produce long period stacking order structure martensites, a triangular morphology associated with the martensitic phase in the Ti-Ni alloy, and a cross-marked morphology associated with the R-phase in the Ti-Ni alloy. Recently, all these self-accommodating morphologies have been well understood; each of them will be explained in order in the following.

There are, for example, 12 crystallographically equivalent lattice correspondences in the DO_3 18R transformation. Because there are six $\{011\}_{DO3}$ planes and in each of them there are two shear directions ($\langle 011\rangle$ and $\langle 01\bar{1}\rangle$) available for producing 18R martensite from the DO_3 structure, the combination of each one of the planes with each one of the shear directions results in 12 crystallographically equivalent lattice correspondences. There are two solutions in phenomenological theoretical calculations for each lattice correspondence, resulting in 24 martensite variants as shown in Fig. 9, where numbers with or without prime indicate the lattice correspondences and label (+) or (-) distinguishes two types of solutions for each lattice correspondence[9].

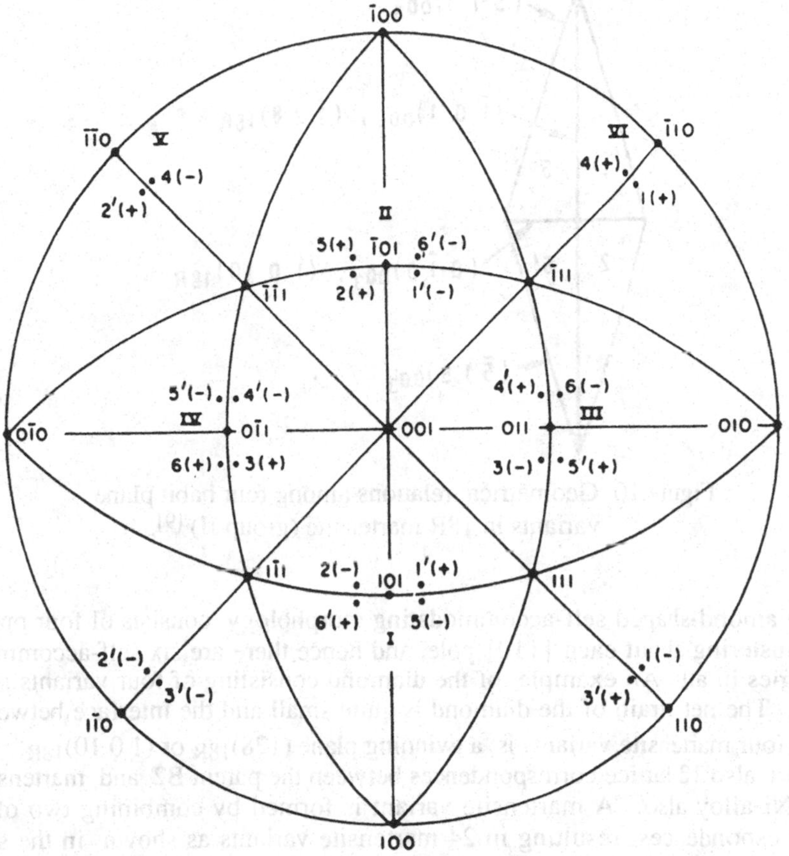

Figure 9 Stereographic projection showing calculated habit plane poles for the 24 martensite(18R) variants of the Cu-Zn-Ga alloy. Four habit plane poles are clustered about each {110} pole. Each of the six {110} groups is designated by the Roman numerals I~VI as shown [9].

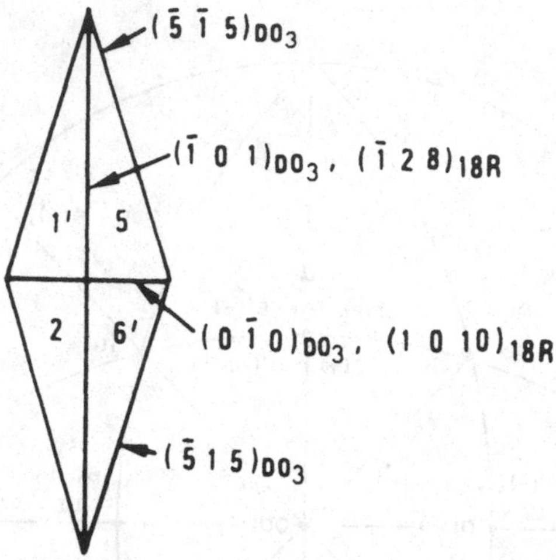

Figure 10 Geometrical relations among four habit plane variants in 18R martensite (group II) [9].

Each diamond-shaped self-accommodating morphology consists of four martensite variants clustering about each {110} pole, and hence there are six self-accommodating morphologies in all. An example of the diamond consisting of four variants is shown in Fig. 10. The net strain of the diamond is quite small and the interface between each two of the four martensite variants is a twinning plane $(\bar{1}28)_{18R}$ or $(1\ 0\ 10)_{18R}$.

There are also 12 lattice correspondences between the parent B2 and martensite B19' in the Ti-Ni alloy also. A martensite variant is formed by combining two of the 12 lattice correspondences, resulting in 24 martensite variants as shown in the standard $(001)_{B2}$ stereographic projection in Figs. 11, where the closed circles indicate the habit planes and the arrows the shear directions[22].

With reference to Fig.11, various possible self-accommodating groups are conceivable. They involve habit planes symmetrically clustered about {001}, {011} and {111} poles relative to the parent phase. However, the trace analysis revealed that the self-accommodating morphology is triangle and consists of 3 of 4 martensite variants clustered around one of three $\{001\}_{B2}$ poles. An example of the triangular morphology is shown schematically in Fig. 12, where the interface between each two of 3 variants corresponds to a twinning plane; i.e. $<011>_M$ Type II twinning plane or $(001)_M$ compound Type I twinning plane[22]. The calculated shape strain matrix of the triangle is quite small, although non-zero. There are 16 groupings which form such a triangular self-accommodating morphology about each of the $\{001\}_{B2}$ poles, resulting in 48 possible combinations in all to form such triangles.

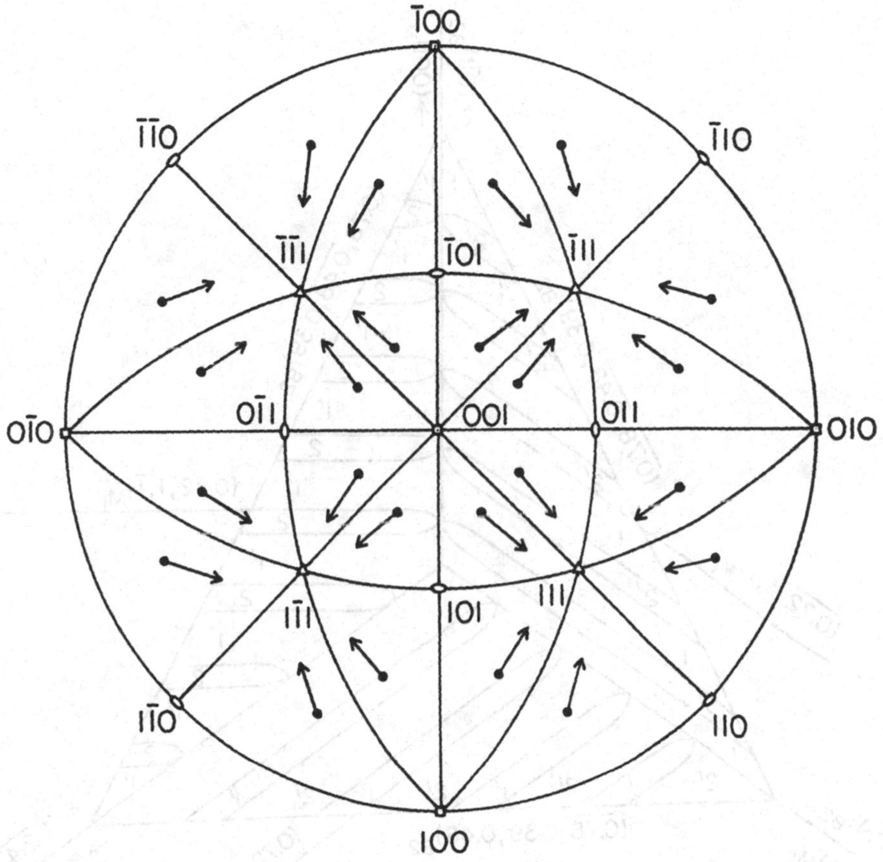

Figure 11 Standard $(001)_{B2}$ stereographic projection showing calculated habit planes (closed circles) and corresponding shape-strain directions (arrows) of the Ti-Ni alloy [22].

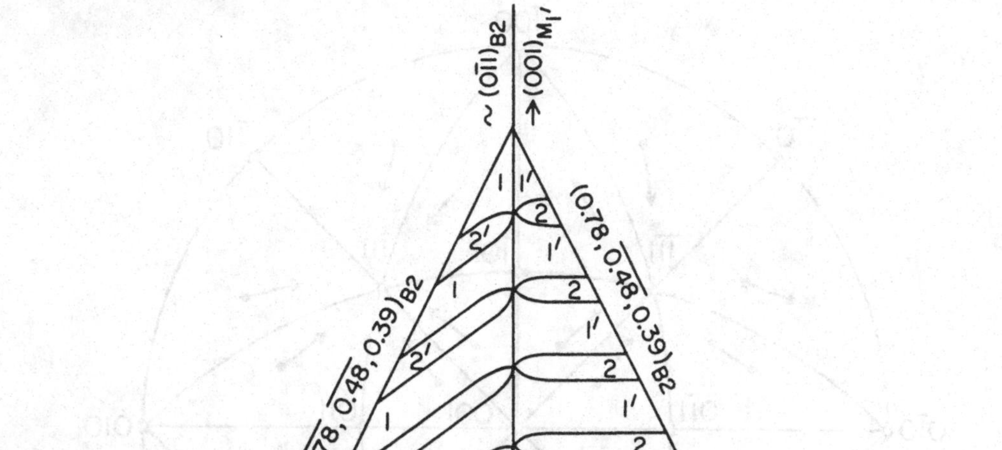

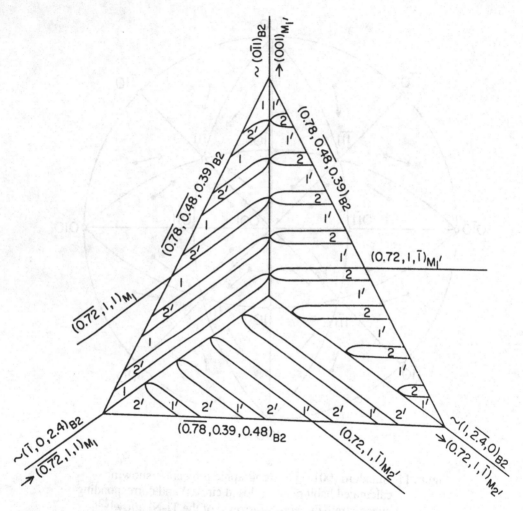

Figure 12 Schematic diagram of triangular B19' martensite morphology depicting crystallographic relationships between variants in the Ti-Ni alloy [22].

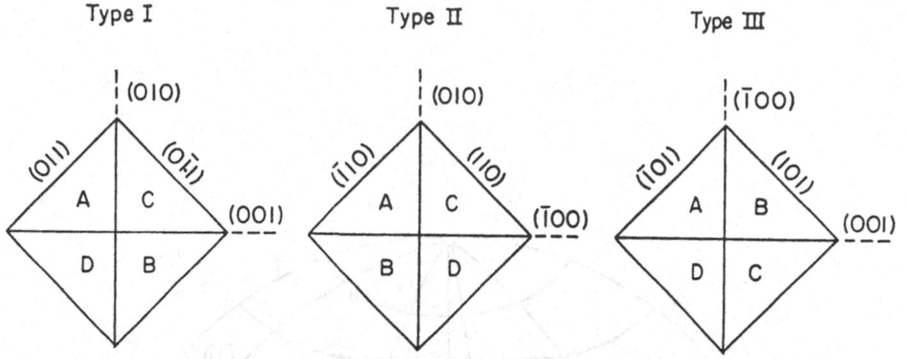

Figure 13 The minimum unit of self-accommodating R-phase variants morphology in the Ti-Ni alloy. Three possible types of variant-combinations are shown [21].

The R-phase transition in the Ti-Ni alloy is essentially a martensitic transformation, and hence the R-phase variants form a typical cross-marked self-accommodating morphology consisting of all 4 types of variants[21]. Three types of the possible minimum unit of self-accommodating morphology are shown in Fig. 13. There are four variants A, B, C and D expressed by the elongated axes $<111>_{B2}$ in the figure, where the twinning planes {011} and {001}, which combine two variants with a twinning relationship, are also indicated. The shape strain matrix of each morphology was nearly the unit matrix.

2.4.2 Variant coalescence upon loading

As explained in the preceding section, each martensite forming a habit plane consists of two lattice-correspondence martensites which have a twinning relationship with each other in β-phase and Ti-Ni alloys, while a single lattice-correspondence R-phase forms a R-phase variant in the Ti-Ni alloy. Besides, each variant is connected with another variant by a twinning plane. These twinning planes move easily at a low stress upon loading, resulting finally in forming the most favorable martensite or R-phase variant. If there are twinning planes remained in the martensite variant, they also move for producing the most favorable lattice-correspondence martensite. At this stage, the maximum recoverable strain is attained. Generally, the maximum recoverable strain is larger than the shape strain calculated by the phenomenological theory.

The maximum recoverable strain can generally be calculated by the lattice distortion (***B***) which generally consists of distortions along the principal axes and shear component parallel to the basal plane of the martensite[8]. Examples of the calculated results for the β_1-β_1' transformation in Cu-Zn-Ga, the B2-B19' transformation in Ti-Ni, and the B2-R transition in Ti-Ni are shown in Figs. 14, 15 and 16, respectively.

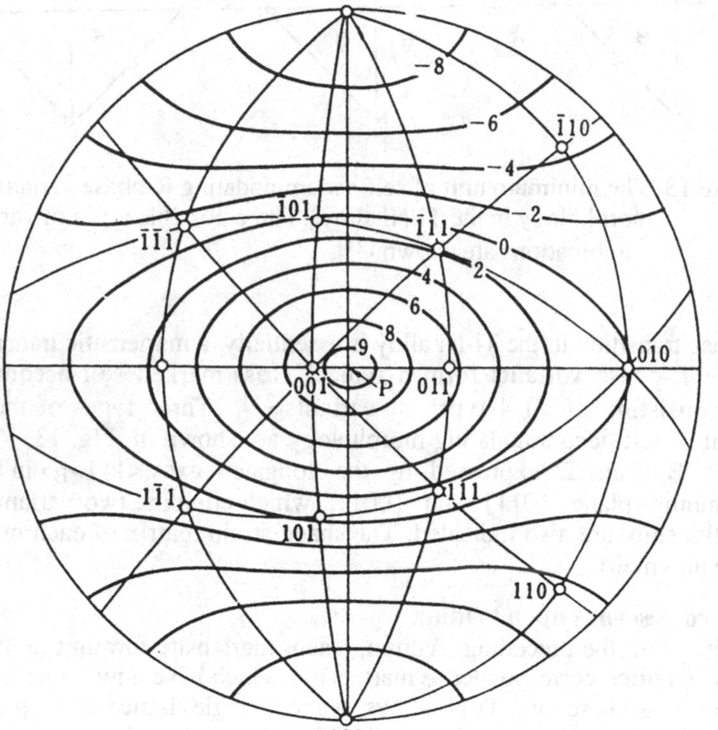

Figure 14 Stereographic representation of the calculated maximum recoverable strain occurring in the β_1 to β_1 (18R) transformation in the Cu-Zn-Ga alloy [9].

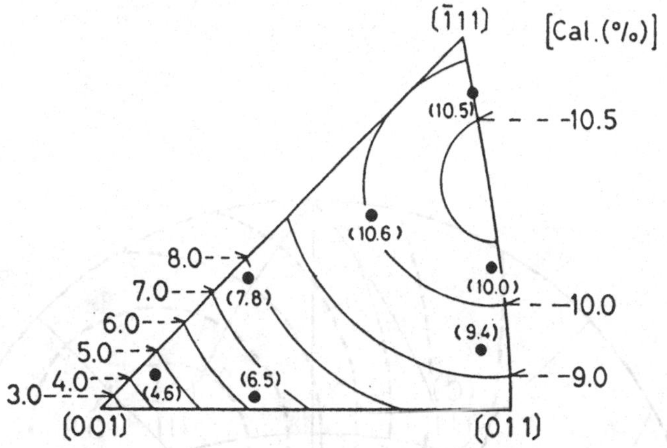

Figure 15 Orientation dependence of the maximum recoverable strain associated with the martensite in the Ti-Ni alloy. Numbers in the parentheses represent the experimental results, and the contour lines the calculated ones [34].

2.4.3 Shape recovery

Shape recovery occurs by the reverse transformation accompanied by the movement of the interface between the parent and martensite phases upon heating. The shape recovery starts at *As* and completes at *Af.* However, the shape recovery occurs in a different manner upon the reverse R-phase transition as shown in Fig. 17, which shows the shape change vs. temperature diagram upon cooling, loading and heating[21]. Although the reverse transition takes place at T_R, most of the strain recovers by the change of unit cell upon heating to T_R. In fact, during heating, no movement of interface between the R-phase and parent phase was observed.

2.5 SUCCESSIVE STRESS-INDUCED TRANSFORMATION

It has been found that stress affects the martensitic transformation not only by changing the transformation temperatures but also by inducing new kinds of martensites. Figure 18 shows stress-strain curves of a Cu-Al-Ni single crystal deformed at various temperatures[11]. In each curve each stage corresponds to a stress-induced martensitic transformation. By plotting the stress for each stage representing a transformation, we can construct a phase diagram representing several martensite phases

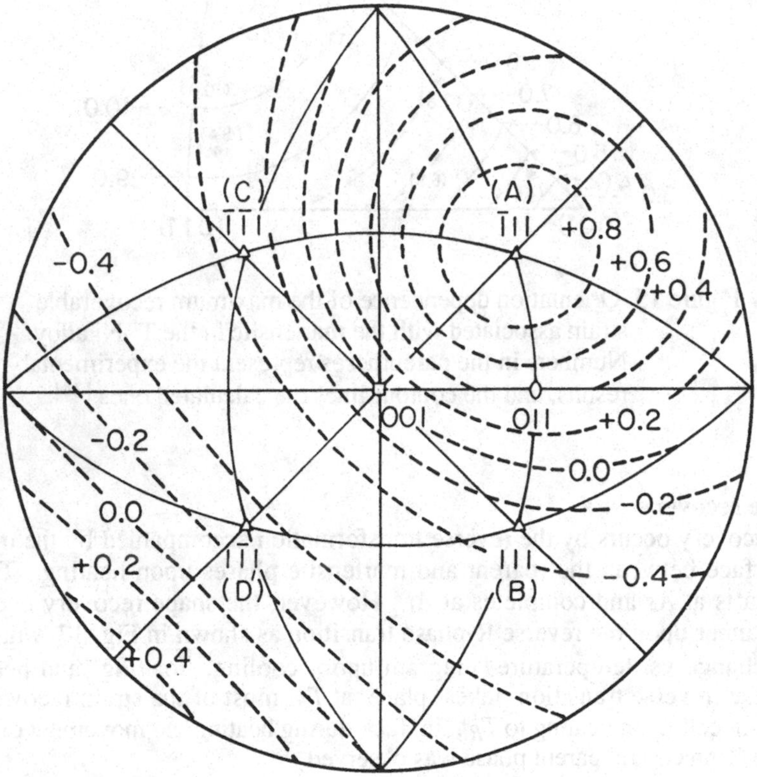

Figure 16 Standard $(001)\beta_2$ projection showing the calculated orientation dependence of the recoverable strain associated with the R-phase in the Ti-Ni alloy [31].

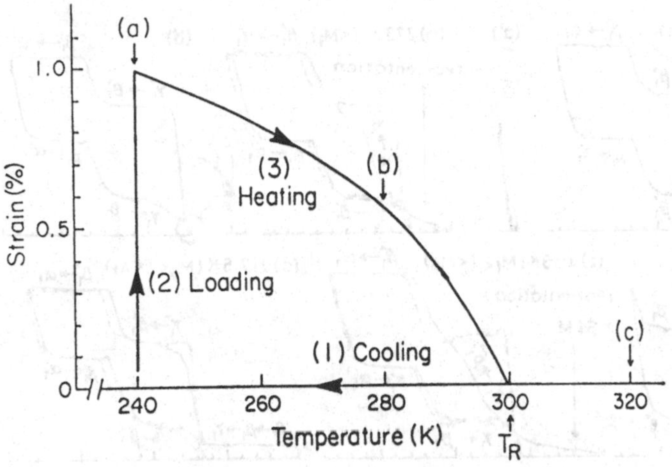

(1) Formation of self-accommodated morphology composed of 4 variants upon cooling
(2) Formation of the most favorable single variant with respect to the applied stress by twinning
(3) Recovery of strain due to the change of rhombohedral angle upon heating

Figure 17 Shape change vs. temperature diagram showing the shape memory effect associated with the R-phase transition in the Ti-Ni alloy [21].

in stress-temperature coordinates as shown in Fig. 19, which is a schematic representation.

There are three types of transformation sequences according to the test temperature. Figures 18(a) and (b) reveal successive transformations upon loading through the path 1 in Fig. 19, i.e. γ_1'-β_1''-α_1' transformation sequence. In this case, the transformation sequence is not crystallographically reversible; α_1'-β_1'-γ_1' transformation sequence appears upon unloading. Figures 18(c) and (d) involve a β_1-γ_1'-β_1''-α_1' transformation sequence. Finally, Fig. 18(e)-(h) involve a β_1=β_1'=α_1' crystallographically reversible transformation sequence.

Ti-Ni alloys also show successive stages of transformation in the stress-strain curve. The deformation is associated with both the R-phase and the martensite in Ti-Ni alloys which include a high density of dislocations and/or fine Ti_3Ni_4 precipitates[30]. Therefore, the deformation behavior is sensitive to test temperature; it is divided into six categories according to the relative relationship between test temperature and transformation temperatures as schematically shown in Fig. 20.

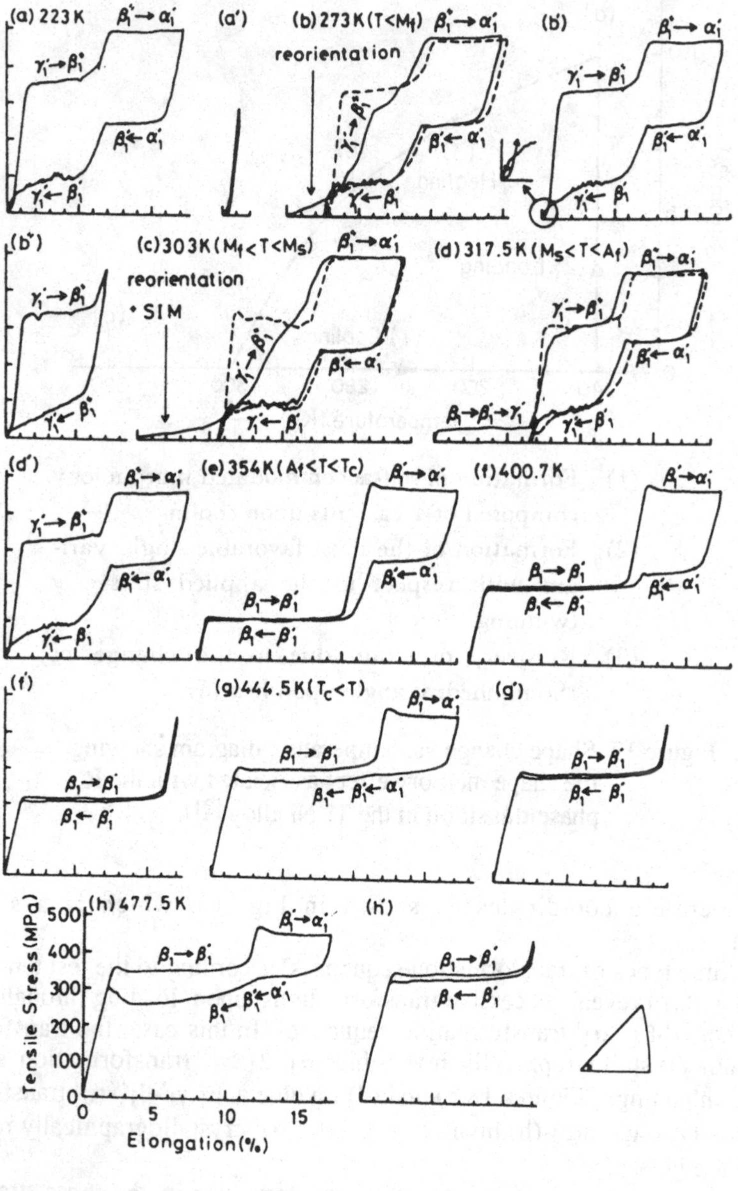

Figure 18 Stress-strain curves as a function of temperature representing successive stress-induced transformations in Cu-14.0Al-4.2Ni (mass%) alloy [11].

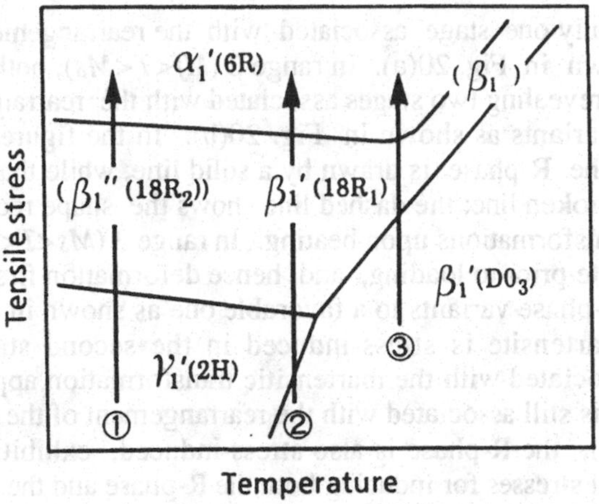

Figure 19 Schematic phase diagram of Cu-Al-Ni alloy in temperature-stress coordinates [11].

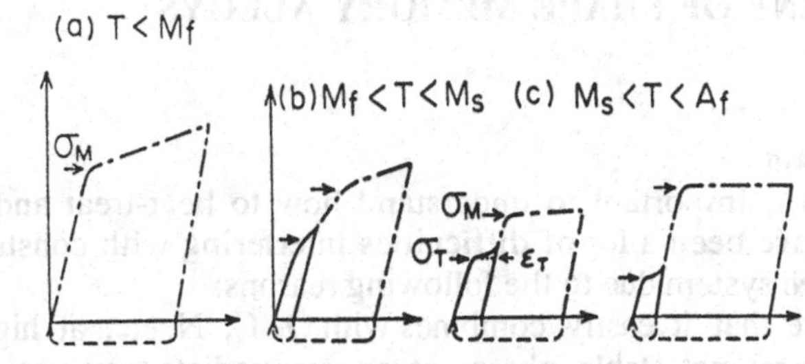

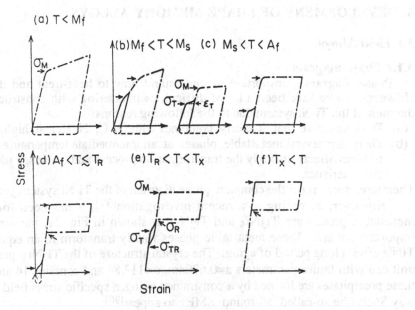

Figure 20 Classification of stress-strain curves according to test temperatures for Ti-Ni alloys exhibiting both the R-phase and martensitic transformations [30].

In range 1 ($T<Mf$), only one stage associated with the rearrangement of martensite variants appears as shown in Fig. 20(a). In range 2 ($Mf<T<Ms$), both the R-phase and the martensite coexist, revealing two stages associated with the rearrangement of the R-phase and martensite variants as shown in Fig. 20(b). In the figure, the stress-strain curve associated with the R-phase is drawn by a solid line, while that associated with the martensite is by a broken line; the dashed line shows the shape recovery associated with the two reverse transformations upon heating. In range 3 ($Ms<T<Af$), the specimen is in a fully R-phase state prior to loading, and hence deformation first proceeds by the rearrangement of the R-phase variants to a favorable one as shown in Fig. 20(c). Upon further loading the martensite is stress-induced in the second stage. In range 4 ($Af<T<T_R$), the PE associated with the martensitic transformation appears, although a part of the deformation is still associated with the rearrangement of the R-phase variants. In range 5 ($T_R<T<Tx$), the R-phase is also stress-induced, exhibiting two-stage PE (Fig. 20(e)). The critical stresses for inducing both the R-phase and the martensite satisfy the Clausius-Clapeyron relationship. Since the slope of the stress-temperature relation for the R-phase is steeper than that for the martensite, both lines cross each other at a temperature Tx. Thus, the deformation associated with the R-phase does not appear in range 6 ($Tx<T$) as shown in Fig. 20(f).

3. DEVELOPMENT OF SHAPE MEMORY ALLOYS

3.1 Ti-Ni Alloys

3.1.1 Phase diagram

Phase diagram is important to understand how to heat-treat and develop alloys. However, there have been a lot of difficulties interfering with constructing the phase diagram of the Ti-Ni system due to the following reasons:

(a) Ti is so active that it easily combines with O, C, N etc. at high temperatures.
(b) There are several metastable phases at an intermediate temperature region and one of them affects strongly the transformation process and the shape memory characteristics.

Therefore, there is not the complete phase diagram of the Ti-Ni system yet at present.

However, according to a recent investigation[37], it has been found that these metastable phases are Ti_3Ni_4 and Ti_2Ni_3 as shown in Fig. 21, the former being the important phase. These metastable phases finally transform to an equilibrium phase $TiNi_3$ after a long period of aging. The crystal structure of the Ti_3Ni_4 precipitates has a unit cell with lattice parameters a=0.670nm, α=113.8^o and contains 14 atoms [38-40]. If these precipitates are formed by a constrained aging, a specific stress field causes a two way SME (the so-called all-round SME) to appear[70].

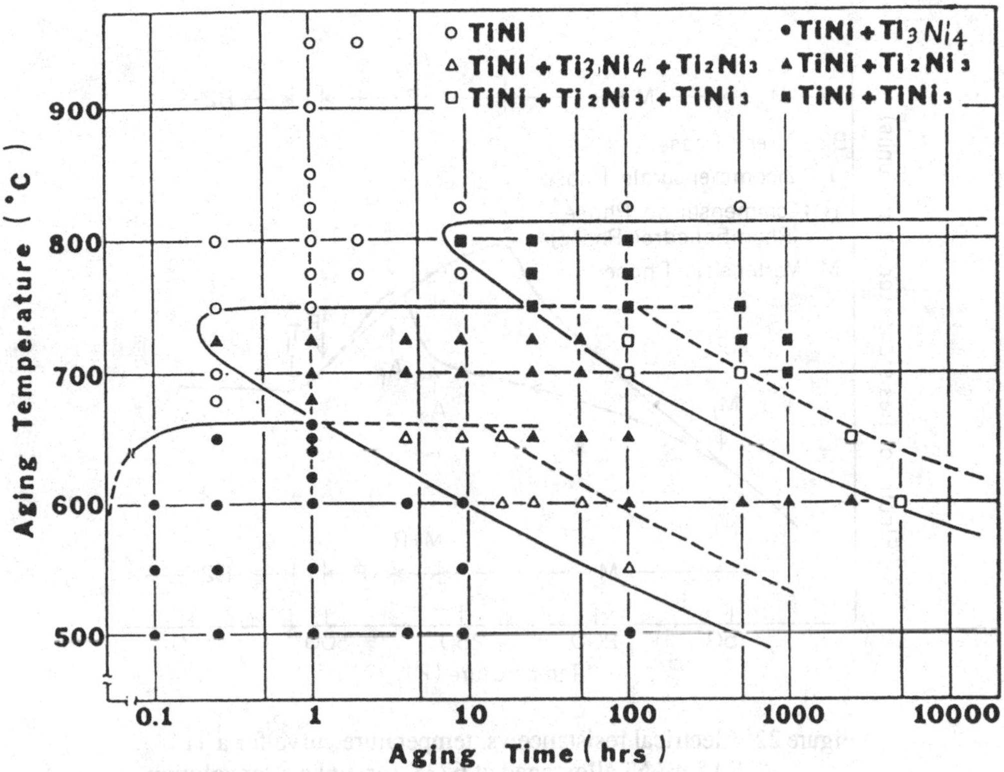

Figure 21 TTT diagram describing aging behavior for Ti-52Ni [37].

3.1.2 Transformation process

One of the causes of the confusion in understanding the martensitic transformation of the Ti-Ni alloy is that so called "premartensitic transition" appears prior to the martensitic transformation. This transition has recently been investigated in the Ti-Ni_{47}-Fe_3(at%) alloy in detail by X-ray diffraction, electron diffraction, electron microscopy and neutron diffraction[24,25]. The Ti-Ni binary alloy also shows the "premartensitic transition", if it is subjected to special thermomechanical treatment which introduces a high density of dislocations and/or fine Ti_3Ni_4 precipitates[30]. The transformation proceeds in the above cases as shown in Fig. 22. The electrical resistance starts to increase on cooling at a critical temperature T_I, and on further cooling the resistance increasing rate becomes rather gradual at a critical temperature T_R which is about 10K lower than T_I. The phase produced first is incommensurate, since the superlattice reflections are deviated from the "one-third position" and the unit cell remains cubic. The incommensurate phase is locked into the commensurate one at T_R. The commensurate phase is called the R-phase,

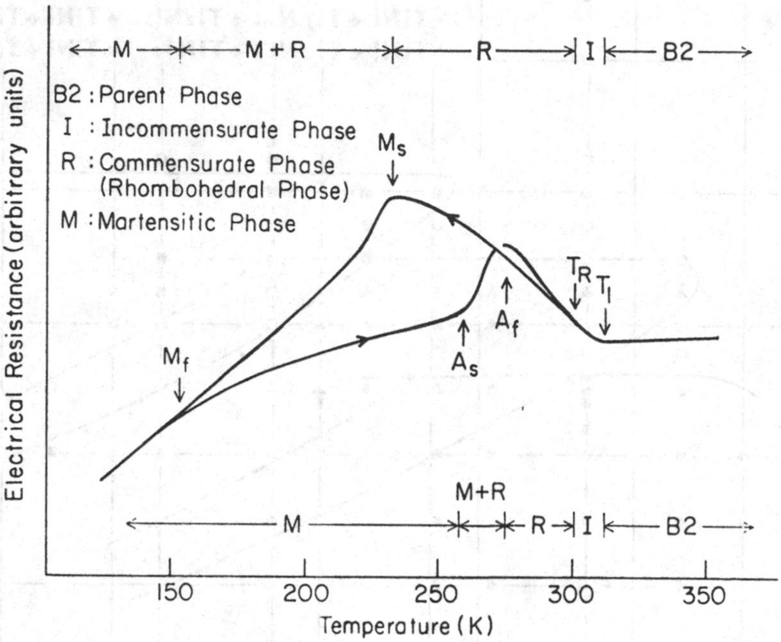

Figure 22 Electrical resistance vs. temperature curve for a Ti-50.5at%Ni alloy aged at 673K for 3.6ks after solution -treatment at 1273K [21].

since the phase is rhombohedrally distorted[28,71]. Thus the transformation in the Ti-Ni alloy containing dislocations and/or precipitates and the Ti-Ni-Fe alloy proceeds as B2→incommensurate phase→R-phase→martensite.

However, solution-treated Ti-Ni binary alloys do not reveal the R-phase transition. A solution-treated equiatomic Ti-Ni alloy shows no electrical resistance increment and exhibits the following transformation sequence upon cooling, i.e. B2-martensite, while a solution-treated Ni-rich Ti-Ni alloy shows an electrical resistance increment at T_I and reveals the following transformation sequence, i.e., B2→incommensurate phase→martensite[30,72].

The R-phase transition usually appears prior to the martensitic transformation when *Ms* is more lowered by some means than *TR*. There are many factors effective to depress *Ms* as follows:

(1) Increasing Ni-content [73,74]
(2) Aging after solution-treatment [13,14,75]
(3) Annealing at temperatures below the recrystallization temperature after cold working [13,75]

(4) Thermal cycling [76,81]
(5) Substitution of a third element [23,75]
Among these factors, factors (2)-(5) are effective to reveal the R-phase transition.

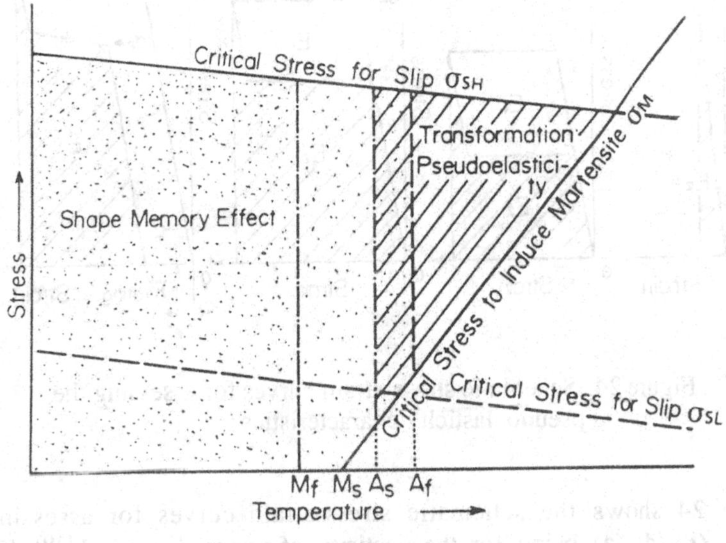

σ_M: The critical stress for inducing martensites
σ_{SL}: The low critical stress for slip
σ_{SH}: The high critical stress for slip.

Figure 23 Schematic diagram representing region of shape memory effect (SME) and pseudoelasticity (PE) in stress-temperature coordinates.

3.1.3 Deformation behavior

The martensitic transformation occurs by loading as well as by cooling. Therefore, martensites are stress-induced even at temperatures above *Ms*. The critical stress (σ_M) for inducing martensites increases with increasing test temperature as shown in Fig. 23, and it satisfies the Clausius-Clapeyron relationship, i.e. a linear relationship between σ_M and temperature[15].

However, if the critical stress (σ_s) for slip is low as indicated by σ_{SL} in the figure, there is no temperature region in which the perfect PE appears. This is because slip deformation always occurs prior to the stress-induced transformation. If σ_s is high as indicated by σ_{SH} in the figure, the perfect PE appears.

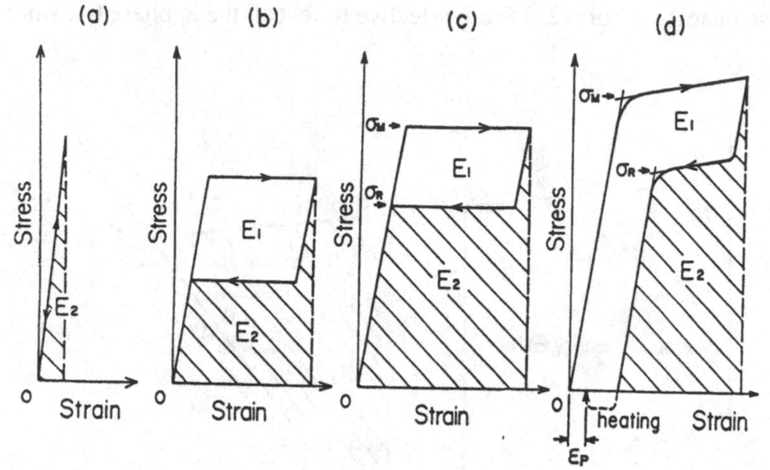

Figure 24 Schematic stress-strain curves for assessing the pseudoelasticity characteristics.

Figure 24 shows the schematic stress-strain curves for assessing the PE characteristics (b)-(d), (a) being for the elasticity of a normal material [19]. The area E_1 is the energy density which is dissipated during one cycle, while E_2 the energy density per unit volume which is stored and available upon unloading. Since the total work done on the specimen is $(E_1 + E_2)$, the efficiency for energy storage is defined as $E_2/(E_1 + E_2)$. This PE characteristics may be applied for storing mechanical energy, because E_2 associated with PE in curve (c) is much larger than that with elasticity in curve (a). For example, E_2 for the Ti-Ni alloy subjected to a special thermo-mechanical treatment amounts to more than 40 times larger than that for an ordinary steel spring[13]. This means that only 2.5 liter of Ti-Ni alloy is enough to store the kinetic energy of an automobile which weighs about one ton and is running with a speed of 50km/h.

By comparing curves (b) and (c), it is easy to understand that both E_2 and η increase with increasing σ_M. However, in case σ_M exceeds σ_s, PE becomes incomplete, since the permanent residual strain (ε_p) is introduced as shown in curve (d). Thus, in order to obtain stable SME and PE characteristics, it is necessary to raise σ_s.

For that purpose, the following three factors are important: (1) annealing temperature, (2) aging temperature, and (3) Ni-concentration[13,14]. Annealing at an intermediate temperature lower than the recrystallization temperature thermally rearranges a high density of dislocations which were introduced by the preceding cold work [13]. Aging Ni-rich alloy produces fine Ti_3Ni_4 ,causing the precipitation hardening. The density of such fine precipitates increases with increasing Ni-concentration, and hence aging effect

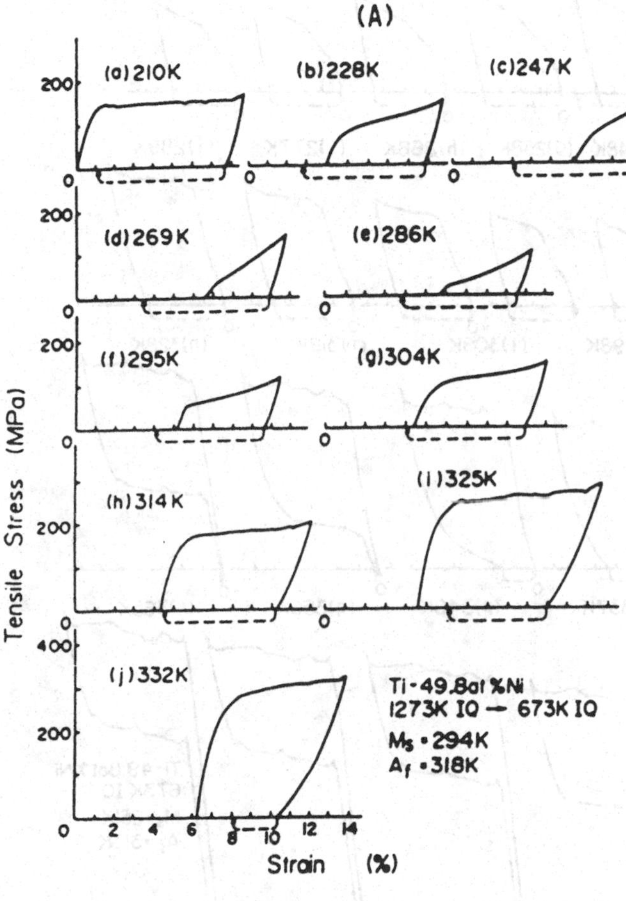

Figure 25(a) Effect of annealing temperature on the stress-strain curves at various test temperatures in the Ti-49.8at%Ni alloys solution-treated at 1273K[13].

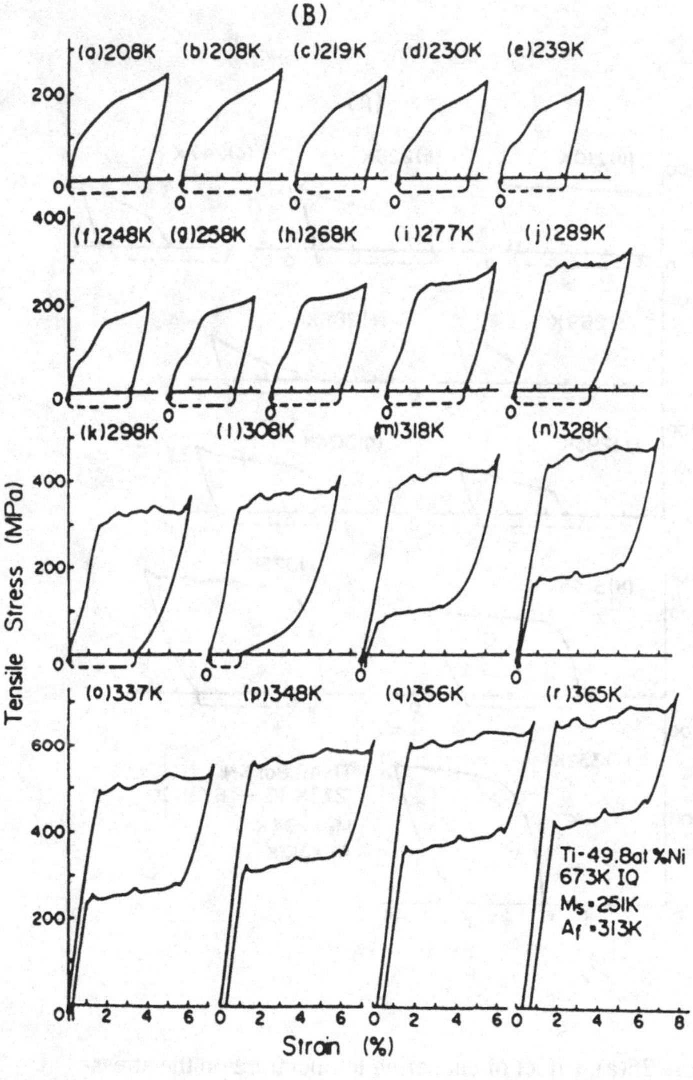

Figure 25(b) Effect of annealing temperature on the stress-strain curves at various test temperatures in the Ti-49.8at%Ni alloys annealed at 673K after cold work[13].

appears more strongly in Ni-rich specimens[13]. These internal structures prevent the movement of dislocations, resulting in raising σ_S.

An example of such effects is shown in Fig. 25 [13], where stress-strain curves obtained at various temperatures are shown for (a) a specimen solution-treated and (b) a specimen annealed at a temperature below the recrystallization temperature after cold work. It is clear that specimen (b) shows excellent SME and PE characteristics when compared to the solution-treated specimen, which shows a large permanent strain.

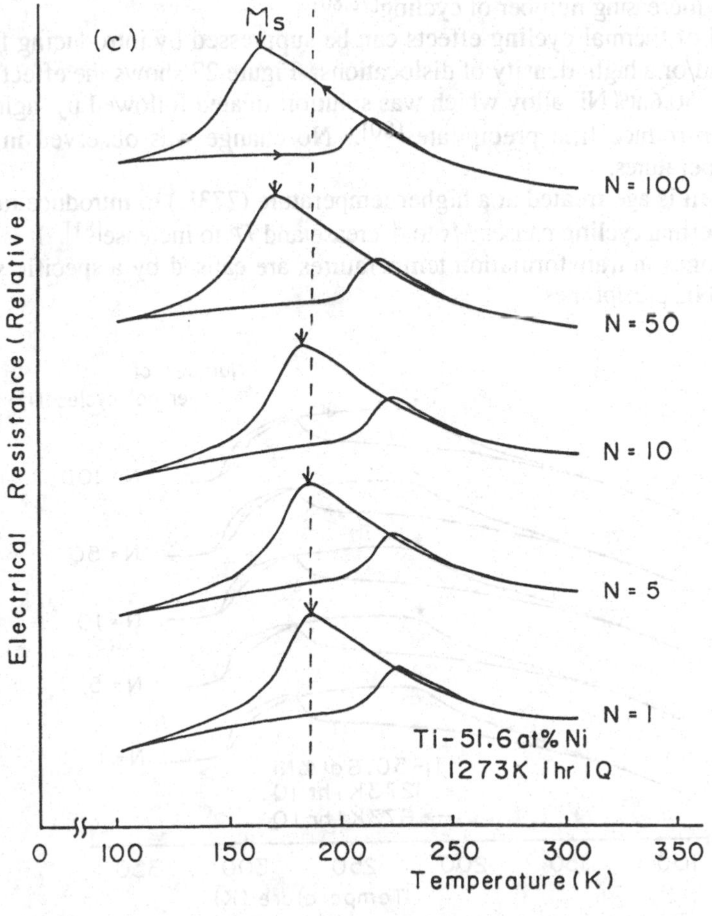

Figure 26 Effect of thermal cycling on the electrical resistance vs. temperature curve for the solution-treated Ti-50.6 at%Ni alloy [80].

3.1.4 Cycling effect

The great advantage of SME and PE is that these functions can be used many times. However, if the alloy is subjected to thermal cycling[76-85] or stress cycling [16,19], the transformation temperatures and deformation behavior will change.

(a) Thermal cycling

Figure 26 shows a typical example of the effect of thermal cycling on the transformation temperatures in the solution-treated Ti-50.6at%Ni alloy[80]. These electrical resistance vs. temperature curves show that *Ms* decreases with increasing number of thermal cycling. Transmission electron microscopy observation revealed that dislocations were introduced by thermal cycling and the density of the dislocations increased with increasing number of cycling[78,80].

Such kind of thermal cycling effects can be suppressed by introducing fine Ti_3Ni_4 precipitates and/or a high density of dislocations. Figure 27 shows the effect of thermal cycling in a Ti-50.6at%Ni alloy which was solution-treated followed by aging at 673K for 3.6ks to introduce fine precipitates[80]. No change was observed in the transformation temperatures.

If a specimen is age-treated at a higher temperature (773K) to introduce rather larger precipitates, thermal cycling causes *Ms* to decrease and *TR* to increase[81]. It is suggested that these changes in transformation temperatures are caused by a specific stress field around the Ti_3Ni_4 precipitates.

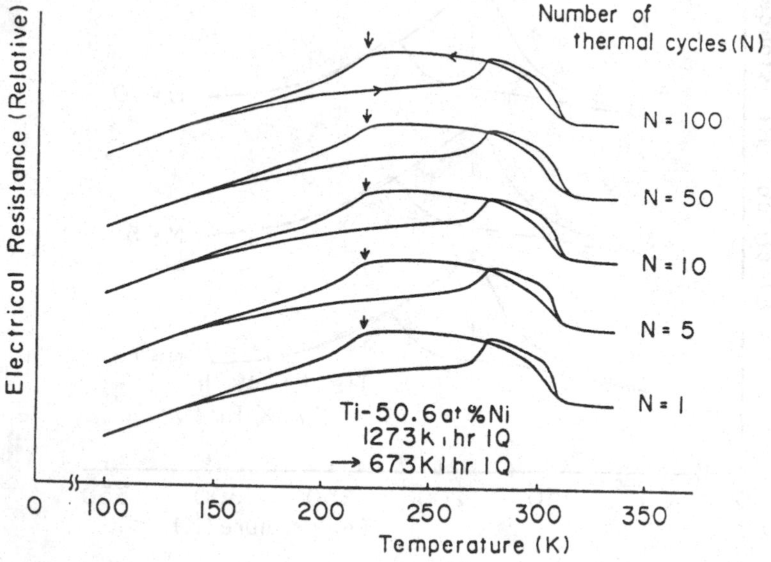

Figure 27 Effect of thermal cycling on the electrical resistance vs. temperature curve for the 50.6at%Ni which was age-treated at 673K after solution-treatment [80].

(b)Stress cycling

Figure 28 shows the effect of stress cycling on the stress-strain curve of the Ti-50.5at%Ni alloy; the stress is controlled by choosing suitable test temperature. General features of the effect of cyclic deformation are as follows:

(1) Residual strain increases as shown by the deviation of the starting point of the stress-strain curve from the original point.

(2) Critical stress σ_M decreases.

(3) Strain or stress hysteresis becomes small.

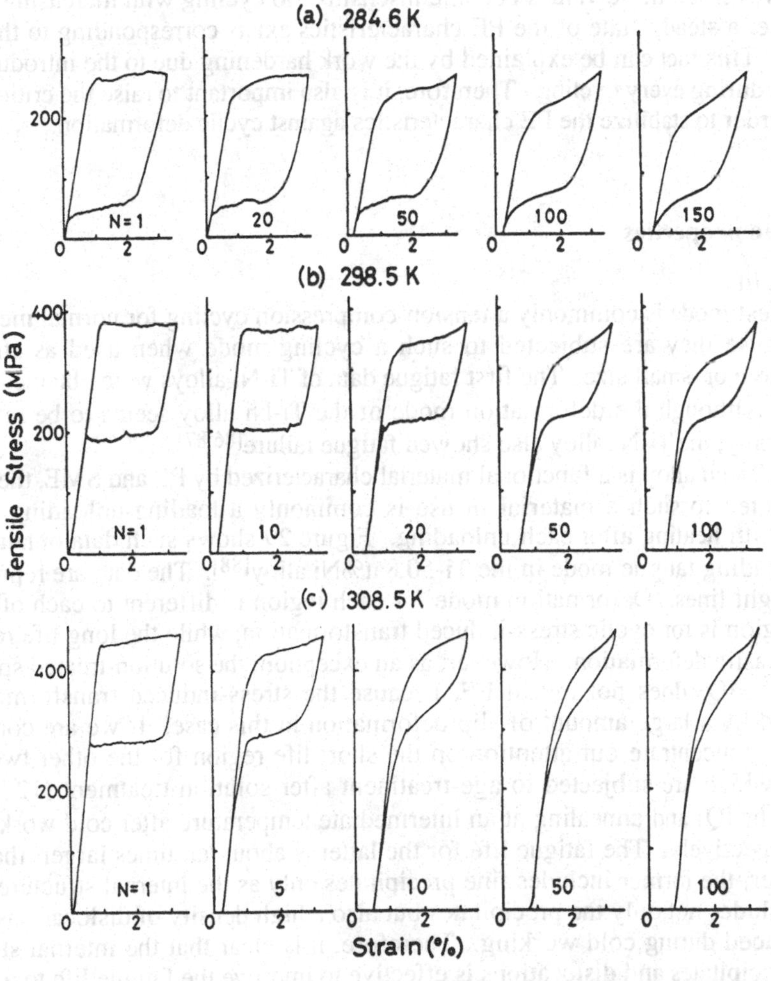

Figure 28 Effect of cyclic deformation on stress-strain curves at various temperatures in the age-treated Ti-50.5at%Ni [19].

The cause for the residual strain is the occurrence of slip deformation during the preceding deformation. Internal stress formed by these slip deformation may assist the formation of the stress-induced martensites; thus σ_M decreases. As the stress-induced transformation occurs like the Luders deformation in stage I [16], σ_M is kept constant for the first cycle. However, since the internal stress field generally has a gradient in its strength, σ_M increases with increasing strain even in stage I after cyclic deformation. As a consequence, the hysteresis becomes smaller by cyclic deformation. The degree of the change in the residual strain, σ_M and the strain hysteresis increase with increasing applied stress. However, all these values become insensitive to cycling with increasing number of cycles, i.e. a steady state of the PE characteristics exists corresponding to the cyclic stress level. This fact can be explained by the work hardening due to the introduction of dislocations during every cycling. Therefore, it is also important to raise the critical stress for slip in order to stabilize the PE characteristics against cyclic deformation.

3.1.5 Fatigue properties

(a) Fatigue life

Fatigue test mode is commonly a tension-compression cycling for normal metals and alloys, because they are subjected to such a cycling mode when used as structural materials, large or small size. The first fatigue data of Ti-Ni alloys were obtained in such a way [86]. Although the deformation mode of the Ti-Ni alloy seems to be reversible macroscopically, the Ti-Ni alloy also showed fatigue failure[86,87].

Since the Ti-Ni alloy is a functional material characterized by PE and SME, the fatigue mode subjected to such a material in use is commonly a loading-unloading cycling without or with heating after each unloading. Figure 29 shows such data obtained in a tension-unloading fatigue mode in the Ti-50.8at%Ni alloy[88]. The data are represented by two straight lines. Deformation mode for each region is different to each other; the short life region is for cyclic stress-induced transformation, while the long life region is for cyclic elastic deformation. However, as an exception, the solution-treated specimen (1273K 1hr IQ) does not reveal PE, because the stress-induced transformation is accompanied by a large amount of slip deformation in this case. If we are concerned with PE, we concentrate our attention on the short life region for the other two types specimens, which are subjected to age-treatment after solution-treatment (1273K 1hr IQ→673K 1hr IQ) and annealing at an intermediate temperature after cold work (673K 1hr IQ), respectively. The fatigue life for the latter is about ten times larger than that for the former; the former includes fine precipitates only as the internal structure, while the latter includes not only the precipitates but also a high density of dislocations which were introduced during cold working. Therefore, it is clear that the internal structure with both precipitates and dislocations is effective to improve the fatigue life to a certain degree. However, it is necessary to investigate the effect of other factors systematically in order to further improve the fatigue life.

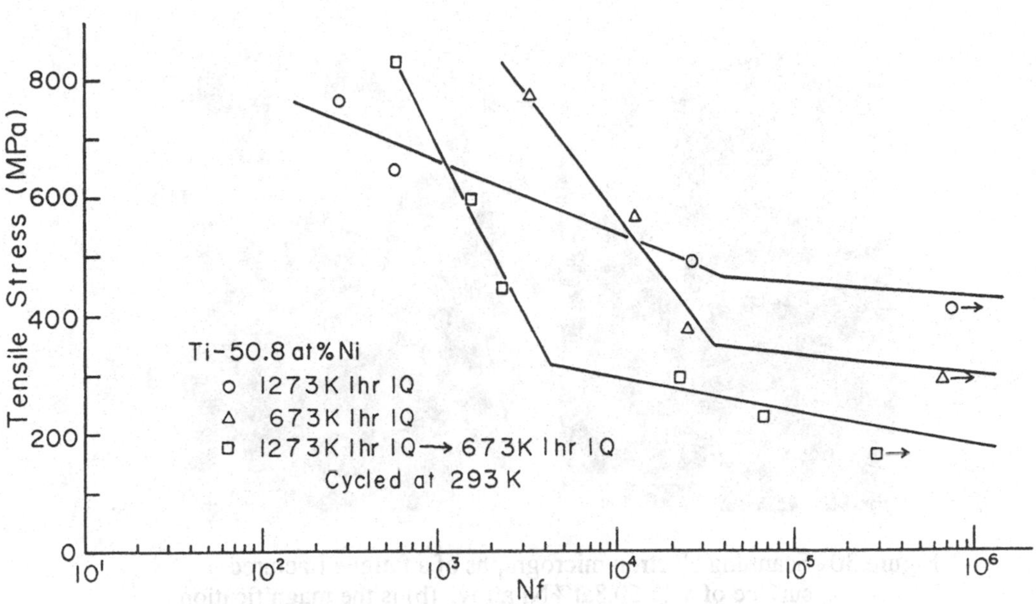

Figure 29 Effect of heat-treatments on the fatigue life of thc Ti-50.8at%Ni alloy which was tested at 293K [88].

(b) Fatigue crack nucleation [89]

The fatigue data mentioned above were obtained using specimens produced in a carbon crucible by a high frequency induction melting method. Therefore, these specimens contained TiC inclusions. Fatigue cracks were frequently observed to nucleate around these inclusions. One of the examples is shown in Fig. 30, in which (a) shows a TiC inclusion at the crack nucleation site, and (b) exhibit the magnified photograph of (a). Therefore, it is expected that purification of alloys lengthen the fatigue life. However, specimens produced by an electron beam melting method showed almost the same fatigue life as that of the specimen containing TiC inclusions.

We now ask why purification of the alloy hardly affects the fatigue life. To answer this question, we need to clarify the fatigue crack nucleation site in specimens produced by an electron beam melting method. Optical micrographic observation revealed that a crack was nucleated at a grain boundary as shown in Fig. 31. It is important to notice that fatigue crack nucleates at about 10% of the total life, irrespective of whether the crack nucleation site is at a TiC inclusion or along a grain boundary. This means that purification of alloys hardly affects the total life.

Since the TiC inclusion does not transform, transformation of the surrounding matrix produces a large strain incompatibility along the interface between the inclusion and the matrix, causing a large stress concentration. In case of a grain boundary, transformation

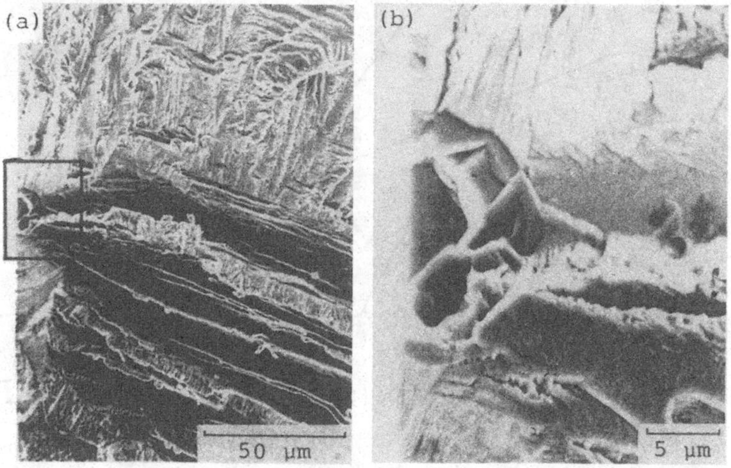

Figure 30 Scanning electron micrographs of a fatigue fractured surface of a Ti-50.8at%Ni alloy. (b) is the magnification of (a) [89].

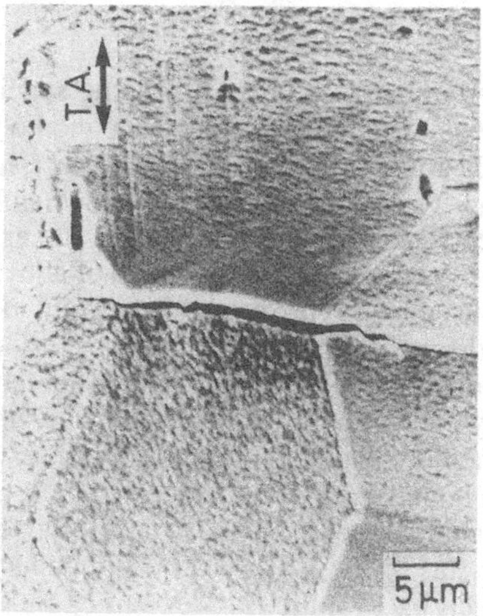

Figure 31 Fatigue crack nucleated along a grain boundary [89].

in one grain causes the second variant to be formed in the neighboring grain in order to make the strain incompatibility along the grain boundary decrease. If the contribution of the second variant is not large enough to reduce the strain incompatibility, a large stress concentration will be created at the grain boundary. Thus both the TiC inclusion and such grain boundary may become crack nucleation sites. However, one hundred percentage of the transformation strain contributes to the strain incompatibility around the interface of the TiC inclusion, while part of the transformation strain will be eliminated from contributing to the strain incompatibility along the grain boundary. This is the reason why a fatigue crack preferentially nucleates at the TiC inclusion rather than along the grain boundary if specimen includes a high density of large TiC inclusions. On the basis of the above discussion, it is possible to improve fatigue life of the Ti-Ni alloy by introducing the combination of purification of alloys and development of a special texture.

(c) Fatigue crack propagation

As mentioned earler, the martensitic transformation is the main deformation mode in SMAs, and this is the essential feature which distinguishes SMAs from the other normal

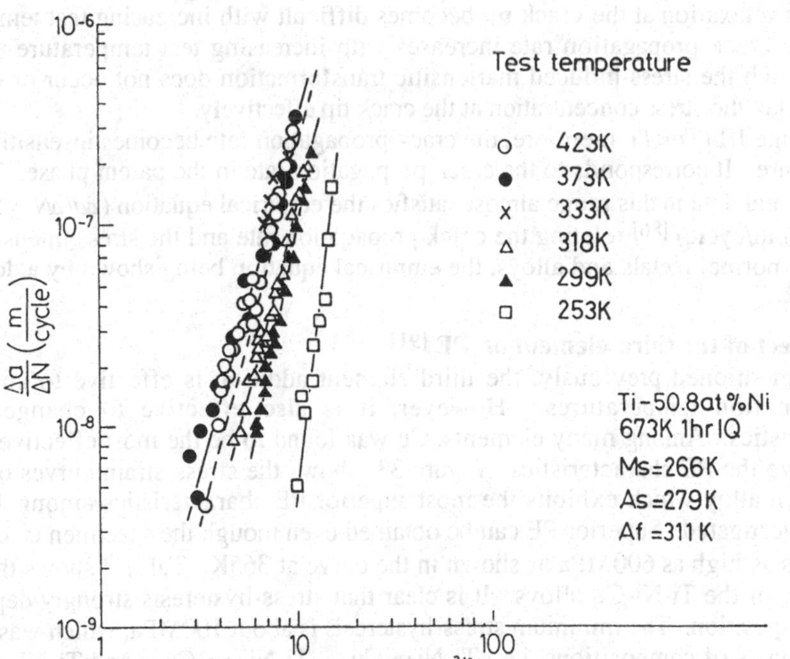

Figure 32 Crack propagation rate as a function of stress intensity factor range for various temperatures in the Ti-50.8at%Ni alloy [90].

metals and alloys. Thus, the martensitic transformation is expected to serve to relax the stress concentration at crack tips in SMAs[86]. Since σ_M is a function of test temperature (*T*), it is expected that the crack propagation rate varies with *T* [90].

Figure 32 shows the relationship between the crack propagation rate ($\Delta a/\Delta N$) and the stress intensity factor range (ΔK) at various temperatures[90]. These data can be divided into three temperature regions as follows. In range I ($T<Ms$) the crack propagation rate shows the minimum value. The martensitic phase will be introduced at a low stress level during the first loading or it may exist prior to loading. In both cases, the martensitic phase remains at a crack tip. Deformation proceeds by the movement of twin boundaries or interfaces between variants in the martensitic phase, and hence the critical stress for the deformation is very low and almost constant irrespective of temperature in this range. Therefore, the stress concentration at the crack tip will be relaxed most effectively in this situation. This is the reason why the crack propagation rate becomes the minimum.

In range II ($Ms<T<Tp$) the crack propagation rate increases in proportion to test temperature. The martensite will be induced at the propagating crack tip every cycling to relax the stress concentration at the crack tip. The critical stress for inducing the martensitic transformation increases with increasing test temperature. This means that the stress relaxation at the crack tip becomes difficult with increasing test temperature. Thus, the crack propagation rate increases with increasing test temperature up to *Tp*, above which the stress-induced martensitic transformation does not occur or σ_M is too high to relax the stress concentration at the crack tip effectively.

In range III ($Tp<T$), therefore, the crack propagation rate becomes insensitive to test temperature. It corresponds to the crack propagation rate in the parent phase. Thus, the experimental data in this range almost satisfies the empirical equation ($\Delta a/\Delta N = 5.1 \times 10^6$ ($\Delta K/Y^{3.5}$) m/cycle) [86] relating the crack propagation rate and the stress intensity factor range for normal metals and alloys, the empirical equation being shown by a dotted line in Fig. 32.

3.1.6 Effect of the third element on PE [91]

As mentioned previously, the third element addition is effective to change the transformation temperatures. However, it is also effective to change the PE characteristics. Among many elements, Cu was found to be the most effective element to improve the PE characteristics. Figure 33 shows the stress-strain curves of the Ti-Ni_{40}-Cu_{10} alloy which exhibits the most superior PE characteristics among Ti-Ni-Cu alloys investigated. Superior PE can be obtained even though the specimen is deformed at a stress as high as 600MPa as shown in the curve at 365K. Table 3 shows the stress-hysteresis of the Ti-Ni-Cu alloys. It is clear that stress-hysteresis strongly depends on alloy composition. The minimum stress-hysteresis is about 100MPa, which was attained in three types of compositions, i.e., Ti-Ni_{40}-Cu_{10}, Ti-$Ni_{44.5}$-Cu_{10} and Ti-$Ni_{45.5}$-Cu_{10}. Therefore, these three alloys can be used as an energy saving material with high efficiency of energy storage.

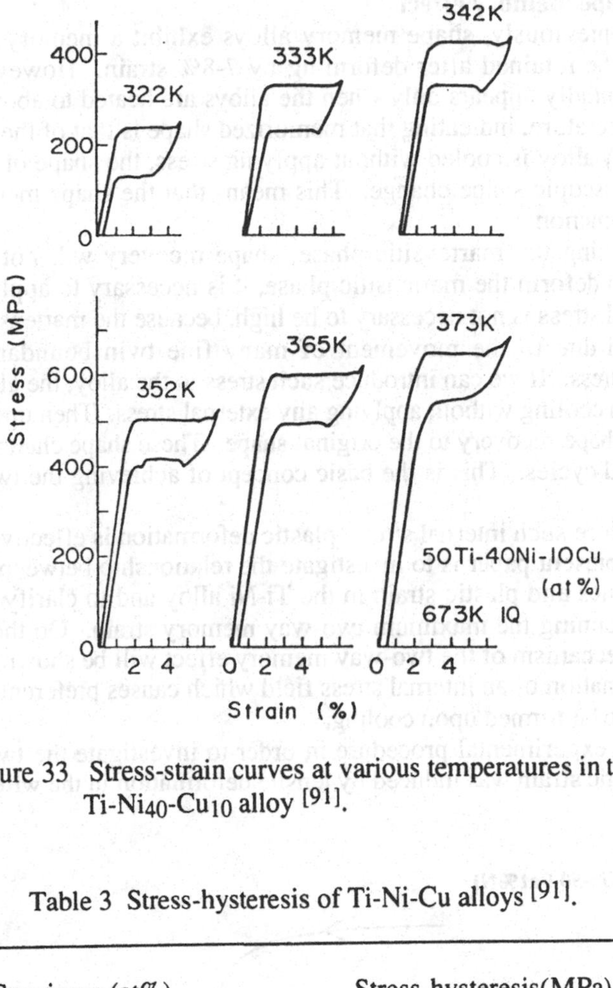

Figure 33 Stress-strain curves at various temperatures in the $Ti-Ni_{40}-Cu_{10}$ alloy [91].

Table 3 Stress-hysteresis of Ti-Ni-Cu alloys [91].

Specimen (at%)	Stress-hysteresis(MPa)
Ti-40.0Ni-10.0Cu	100
Ti-44.5Ni-10.0Cu	100
Ti-45.5Ni-10.0Cu	100
Ti-48.5Ni-10.0Cu	400
Ti-49.5Ni-5.0Cu	300
Ti-50.5Ni-5.0Cu	200

3.1.7 Two-way shape memory effect

As mentioned previously, shape memory alloys exhibit a memory such that the original shape can be retained after deforming by 7-8% strain. However, the shape memory effect essentially appears only when the alloys are heated to above the reverse transformation temperature, indicating that memorized shape is that of the parent phase. If the shape memory alloy is cooled without applying stress, the shape of the alloy will not show any macroscopic shape change. This means that the shape memory effect is one-way type phenomenon.

Without deforming the martensitic phase, shape recovery will not appear upon heating. In order to deform the martensitic phase, it is necessary to apply force to the alloys. Such applied stress is not necessary to be high, because the martensitic phase can be easily deformed due to the movement of many fine twin boundaries under an application of low stress. If we can introduce such stress in the alloy, the alloy will show a shape change upon cooling without applying any external stress. Then upon heating the alloy will exhibit a shape recovery to the original shape. These shape changes will repeat upon further thermal cycles. This is the basic concept of achieving the two-way shape memory effect.

In order to introduce such internal stress, plastic deformation is effectively available. The purpose of the present paper is to investigate the relationship between the two-way memory characteristics and plastic strain in the Ti-Ni alloy and to clarify the optimum plastic strain for obtaining the maximum two-way memory strain. On the basis of the above results, the mechanism of the two-way memory effect will be shown by clarifying the origin of the formation of an internal stress field which causes preferentially oriented martensite variants to be formed upon cooling.

Following is the experimental procedure in order to investigate the two-way shape memory effect. Plastic strain was induced by tensile deformation in the wire specimens

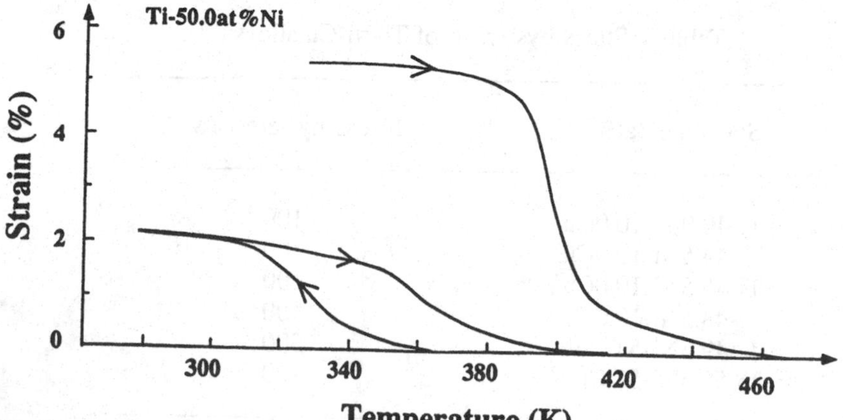

Figure 34 Strain vs. temperature curve exhibiting the two-way memory

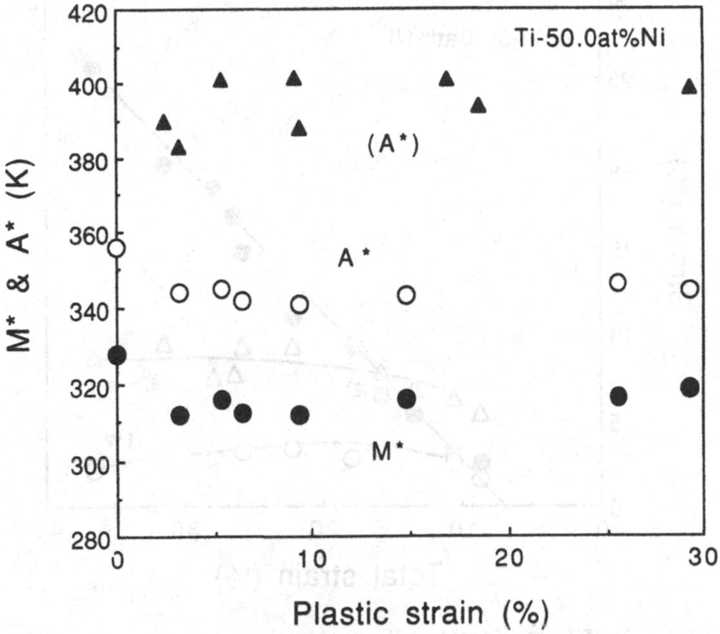

Figure 35 Effect of plastic strain on trans-formation temperatures

with gauge length of 20 mm. After removing the applied stress, the specimens were heated under a constant stress of 3MPa, which is low enough not to disturb the intrinsic deformation behavior of the specimens and high enough to control the movement of the crosshead of the tensile machine during testing. Then the specimens were cooled down to below *Mf* in order to measure the two-way memory strain. The transformation temperatures were determined in specimens with or without prestraining using DSC.

Figure 34 shows an example of strain vs. temperature relationship exhibiting the two-way memory behavior in a predeformed Ti-50.0at%Ni, the prestrain being 2.5%. After unloading, the specimen was first heated to a temperature region above Af of an unprestrained specimen. Large shape recovery occurred due to the reverse martensitic transformation. After finishing the shape recovery, the specimen was cooled. The specimen gradually elongated up to about 2% strain due to the martensitic transformation. Then the specimen was again heated, resulting in shape recovery associated with the reverse transformation. However, the reverse transformation temperature was lower than that of the first heating. This two-way memory effect can be repeated without changing the strain vs. temperature curve.

In order to measure the transformation temperatures of a specimen which was predeformed, the specimen was spark-cut into 4 mm length without strain. Figure 35

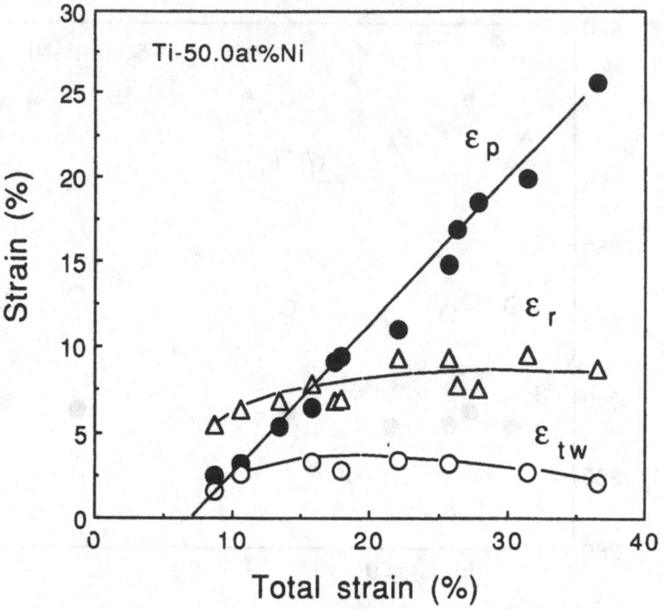

Figure 36 Effect of total strain on plastic strain, recovery strain and two-way memory strain

shows the first reverse transformation peak temperature (A^*), the martensitic transformation peak temperature M^* and the second reverse transformation peak temperature A^*. They are indicated by closed triangles, closed circles and open circles, respectively. (A^*) increases with increasing plastic strain , while M^* and A^* decrease. However, all these transformation temperatures become constant irrespective of plastic strain when the plastic strain is higher than 15%.

Figure 36 shows the relationship between various types of strains and total applied strain. The plastic strain ε_p increases lineally with total strain. The recovery strain ε_r upon the first heating and the two-way memory strain ε_{tw} are also shown, ε_r being always higher than ε_{tw}.

The two-way memory strain ε_{tw} is shown as a function of plastic strain in Fig. 37. The two-way memory strain increases with increasing plastic strain until getting the maximum value at about 10% plastic strain. Then the two-way memory strain starts to decrease. This result shows that there is an optimum value of plastic strain for achieving the maximum two-way memory strain.

In the above, specimens were prestrained at at Ms. Since the effect of prestrain is considered to depend on deformation temperature, Ti-51.0at%Ni were prestrained at

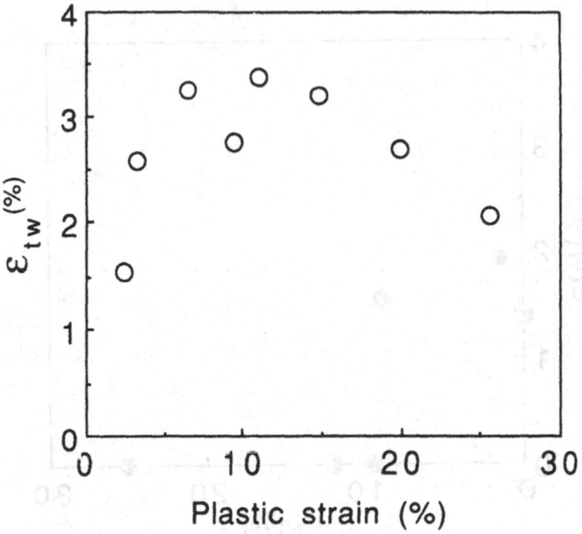

Figure 37 Effect of plastic strain on two-way memory strain

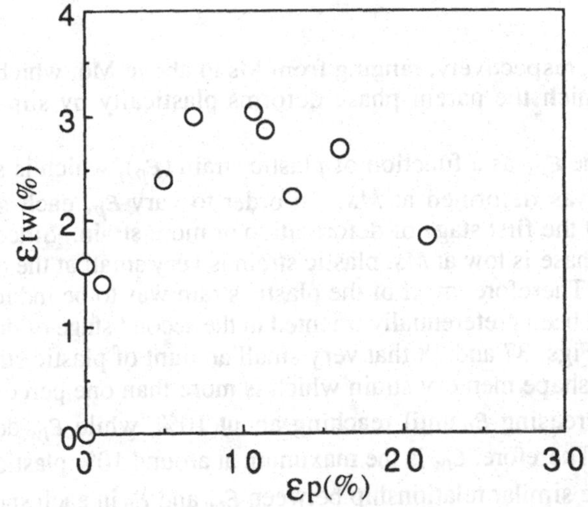

Figure 38 Two-way shape memory strain vs. plastic strain in Ti-51.0at%Ni specimens which were prestrained at Ms.

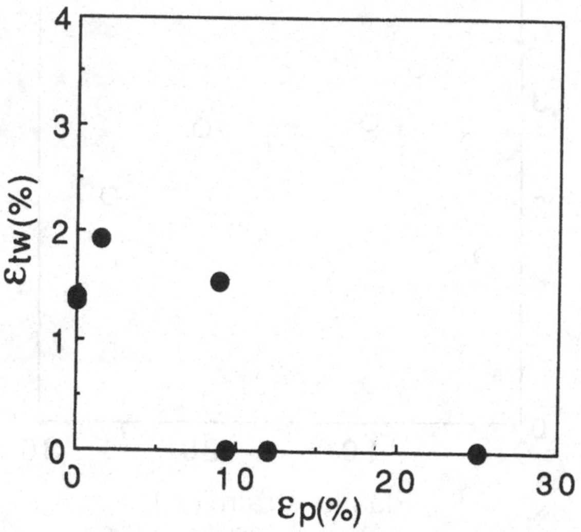

Figure 39 Two-way shape memory strain vs. plastic strain in Ti-51.0at%Ni specimens which were prestrained till the end of the first stage of deformation at various temperatures.

different temperatures, respectively, ranging from Ms to above Md, which is the critical temperature above which the parent phase deforms plastically by slip instead of by transformation.

Figure 38 shows the ε_{tw} as a function of plastic strain (ε_p), which is similar to Fig. 37. Each specimen was deformed at Ms. In order to vary ε_p, each specimen was deformed to the end of the first stage of deformation or more strain. Since the stress for inducing martensitic phase is low at Ms, plastic strain is very small at the end of the first stage of deformation. Therefore, most of the plastic strain was to be induced after most martensite variants had been preferentially oriented in the second stage of deformation.

It is first found in Figs. 37 and 38 that very small amount of plastic strain is enough for inducing two-way shape memory strain which is more than one percent. The strain ε_{tw} increases with increasing ε_p until reaching about 10%, while ε_{tw} decreases with further increasing ε_p. Therefore, ε_{tw} is the maximum at around 10% plastic strain.

Figure 39 shows the similar relationship between ε_{tw} and ε_p in each specimen which was deformed till the end of the first stage of deformation at different temperatures. In this case, the amount of plastic strain was varied by changing deformation temperature.

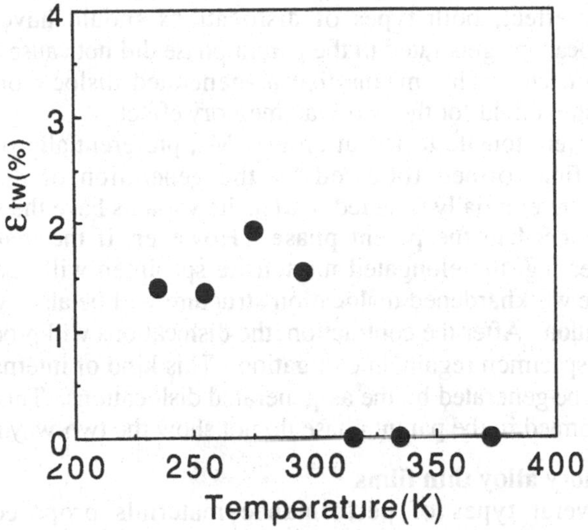

Figure 40 Two-way shape memory strain vs. deformation temperature in Ti-51.0at%Ni specimens which were prestrained till the end of the first stage of deformation at various temperatures.

The two-way shape memory effect was observed in a plastic strain range of less than 10%. However, it was not observed when the plastic strain becomes more than 10%. This is very different from the results in Fig.38, where ε_{tw} was maximum at 10% plastic strain.

The data in Figure 39 were replotted as a function of deformation temperature in Figure 40. The strain ε_{tw} increases with increasing deformation temperature till about 280K. However, it started to decrease rapidly with further increasing deformation temperature, and finally became zero in a temperature range of above about 300K. It was observed that specimens deformed plastically in B2 phase when the temperature is higher than 300K; this was confirmed by the facts that the stress for inducing the plastic deformation deviated from the stress-temperature line which was predicted by the Clausius-Clapeyron relationship and that the stress for the plastic deformation was almost constant irrespective of temperature above 300K. However, the specimens deformed plastically in the martensitic phase when they were deformed below 280K. In the intermediate temperature region, i.e., between 280K and 300K, plastic deformation seemed to occur both in the B2 and martensite phases.

In the above section, it was pointed out that there are two types of dislocations; i.e., dislocations generated in the parent phase and the martensite phase, respectively. If the internal stress which were induced due to dislocations themselves is the cause for the two-way memory effect, both types of dislocations should have the same effect. However, the dislocations generated in the parent phase did not cause the two-way shape memory effect to occur. This means that as-generated dislocations will not induce effective internal stress field for the two-way memory effect.

If specimens were tensile tested at around Ms, preferentially oriented martensite variants will be first formed followed by the generation of dislocations. Such dislocations in the preferentially oriented martensite variants have the same nature as that of dislocations induced in the parent phase. However, if the martensite is reverse transformed by heating, the elongated martensite specimen will contract with a large driving force. The workhardened dislocation structure will be also contracted upon the reverse transformation. After the contraction, the dislocations will produce a spring-back force to make the specimen regain an elongation. This kind of internal stress or spring-back force can not be generated by the as-generated dislocations. This is the reason why the specimens deformed in the parent phase do not show the two-way memory effect.

3.1.8 Shape memory alloy thin films

Among several types of performance materials proposed for fabricating microactuators, Ti-Ni shape memory alloy has great advantages such as large deformation and recovery force; besides, even an intrinsic disadvantage of the shape memory alloy such as a slow response due to the limitation of cooling rate can be greatly improved in thin films, because the surface/volume ratio of materials will drastically increases in micro size actuators.

Therefore, the development of Ti-Ni shape memory thin films has been demanded in the field of micromachines, and several efforts have been made in order to fabricate Ti-Ni thin films using a sputter-deposition method. Some of these results showed that the micromachining technology is useful for making microstructures consisting of a silicon substrate and a Ti-Ni thin film. However, the shape memory behavior of Ti-Ni thin films has not been characterized by mechanical tests, because in order to characterize the shape memory effect it is necessary to apply stress on the films. If the films contain micro defects which are characteristic of sputter-deposited films and other elements such as oxygen and hydrogen, they becomes brittle. Application of stress to such films will cause fracture. Without information about the characteristics of materials, any trial for the improvement in sputtering process will be ineffective. Therefore, it is very important both to establish a fabrication method of a high quality thin films which endure a high stress and to develop mechanical testing methods for evaluating shape memory and superelasticity characteristics of thin films.

Very recently, mechanical behavior of Ti-Ni thin films has been characterized by tensile tests [92-96]. In 1992 the present author and his group, using a thermomechanical testing, succeeded in characterizing a perfect shape memory effect associated with both the R-phase and martensitic transformations in sputter-deposited Ti-Ni thin films[92]. They also determined the crystal structures of the parent and martensitic phases to be B2 and monoclinic, respectively, by X-ray diffractometry. The shape memory characteristics of thin films strongly depend on metallurgical factors and sputtering

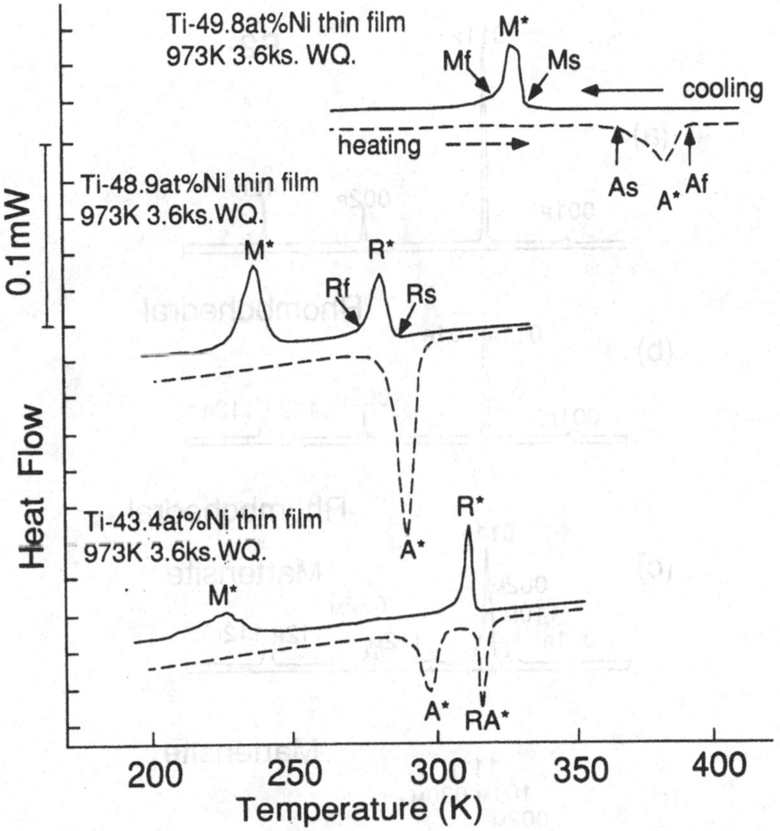

Figure 41 DSC curves and transformation temperatures of Ti-Ni thin films

conditions. The former includes alloy composition, annealing temperature, aging temperature, while the latter includes Ar pressure, sputtering power, substrate temperature and so forth. In the following, some information about the mechanical and transformation properties of sputter-deposited Ti-Ni thin films will be shown, i.e, an aging effect on transformation temperatures, characterization of shape memory behavior,

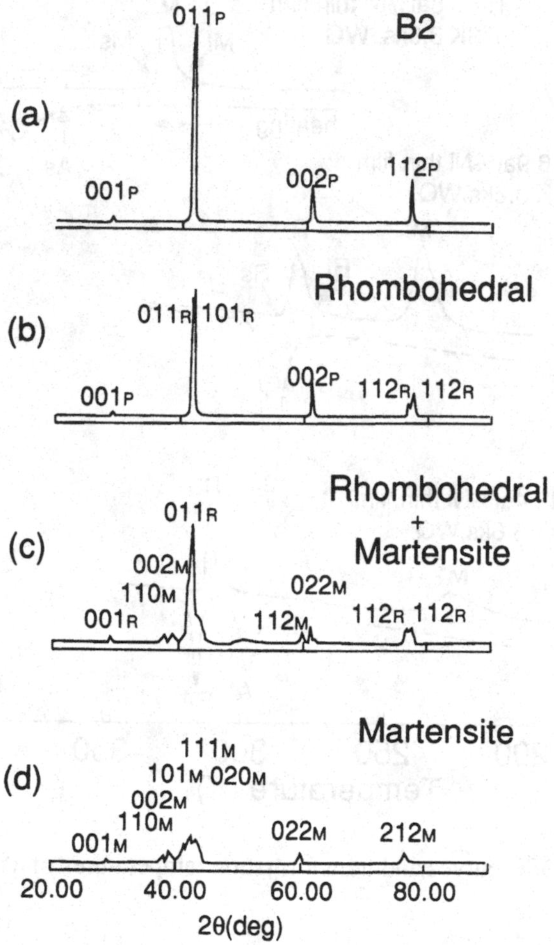

42 X-ray diffraction profiles obtained from B2, R-phase and martensitic phase

and successive observation of the crystal structural change among the parent phase, R-phase and martensitic phase.

Since as-sputtered films were amorphous, all films were heat-treated above the crystallization temperature, which was determined to be 756K by DSC. Figure 41 shows results of DSC measurement in Ti-Ni thin films with different Ni-contents which were annealed at 973K for 3.6ks and followed by quenching into water. Three different types of DSC curves are shown. They are characterized as follows: (1) Ti-49.8at%Ni film shows a single peak both upon cooling (solid line) and heating (dashed line). The martensitic transformation occurs from the parent phase (B2 structure) to the martensite (monoclinic structure) at 327K(peak temperature M^*) upon cooling, while the reverse martensitic transformation occurs from the martensite to the parent phase at 381K(peak temperature A^*). It is important to notice that both M^* and A^* are located above room temperature (RT), because microactuators made of this film work without using any cooling devise. (2) Ti-48.9at%Ni film shows double peaks upon cooling. The first peak at 280K(peak temperature R^*) corresponds to the R-phase (Rhombohedral phase) transformation from the parent phase to the R-phase, while the second peak at 227K(M^*) to the martensitic transformation from the R-phase to the martensite. Upon heating, a single peak appears at 288K(A^*) due to the reverse martensitic transformation from the martensite directly to the parent phase. In this case, a cooling devise is necessary in order to use an actuator function, because the transformation temperatures are lower than RT. (3) Ti-43.4at%Ni film shows double peaks both upon cooling and heating. The meanings of the two peaks at 310K(R^*) and 223K(M^*) upon cooling are the same as the above case. Upon heating, the first peak at 297K(A^*) corresponds to the reverse martensitic transformation from the martensite to the R-phase, while the second peak at 314K(AR^*) to the reverse R-phase transformation from the R-phase to the parent phase. Since M^* and A^* are lower than RT, the martensitic transformation is not available for an actuator without a cooling device. However, the R-phase transformation is available without using any cooling devise, because R^* and AR^* are higher than RT. The R-phase transformation is characterized by a small thermal hysteresis; i.e., the temperature difference between R^* and AR^* is 4K. Quick response is expected for the movement of an actuator using such R-phase transformation.

Figure 42 shows X-ray diffraction patterns obtained from different phases in a Ti-Ni film which shows a two-stage transformation upon cooling similarly to the second or third case in Fig. 41. Diffraction pattern (a) observed at a temperature above R^* indicates that the parent phase is of B2 structure with a lattice constant of 0.30182nm.

Diffraction peaks $(011)_P$ and $(112)_P$ split into two peaks at a temperature between R^* and Ms, which is located at the right tail of the DSC peak M^*, as shown in Fig. 42(b). This indicates that the structure of the film is R-phase in the temperature region. The lattice constants of the R-phase depend on temperature; i.e., the rhombohedral angle a of the R-phase structure decreases with decreasing temperature below the R-phase transformation start temperature Rs as shown in Fig. 43, while the lattice length is kept

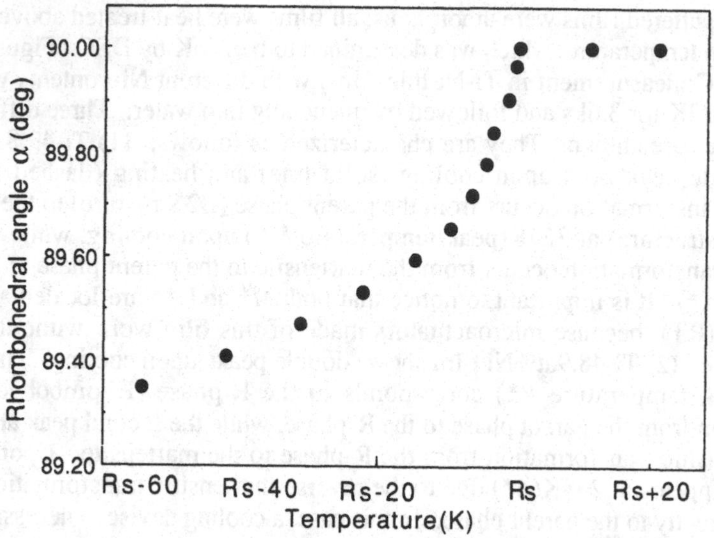

Figure 43 Rhombohedral angle as a function of temperature

constant irrespective of temperature and the same as that of the parent phase. This implies that the strain induced due to the R-phase increases with decreasing temperature. The R-phase starts to transform into the martensitic phase at *Ms* and finish the transformation at the left tail (*Mf*) of the DSC peak *M**.

Diffraction pattern (c) was obtained at a temperature between *Ms* and *Mf*, so that it consists of diffraction peaks of two phases, i.e., the R-phase and the martensitic phase. Diffraction pattern (d) obtained at a temperature below *Mf* shows that the film is fully in the martensitic phase. The martensite is of a monoclinic structure with the following lattice constants; i.e., a = 0.29166nm, b = 0.41475nm, c = 0.46515 and $\beta = 97.167^{\circ}$.

Controlling transformation temperatures is one of the important techniques in order to fabricate shape memory alloy thin film microactuators suitable for various purposes. There are several factors with which the transformation temperatures can be varied; i.e., Ni-content, third element, annealing, aging, and so on. Among these factors, aging effect is shown in Fig. 44. Three types of transformation temperatures, *Rs*, *As* and *Ms*, of Ti-51.9at%Ni thin film are shown as a function of aging time; the films were solution-

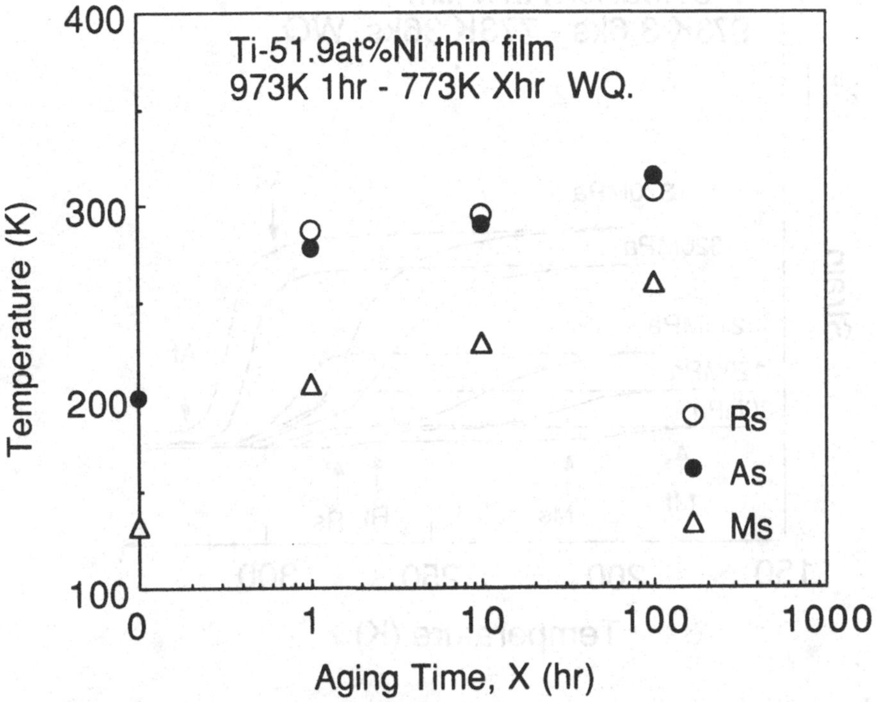

Figure 44 Effect of aging time on transformation temperatures of Ti-51.9at%Ni thin film

treated at 973K for 1 hour (3.6ks) and followed by aging at 773K for various times. As-solution-treated film indicated that *Ms* and *As* are 130K and 202K, respectively. It did not show clear R-phase transformation. However, the film revealed the R-phase transformation at *Rs* (288K) after aging it at 773K for 3.6ks; *Ms* and *As* increased by the aging. These transformation temperatures increased further with increasing aging time. Since *Ms* and *As* are 258K and 316K, respectively, after aging for 100 hours (360ks), it is understood that aging is effective in varying transformation temperatures more than 100K.

The above data such as crystal structures and transformation temperatures are measurable even in Ti-Ni thin films which include microdefects and contamination.

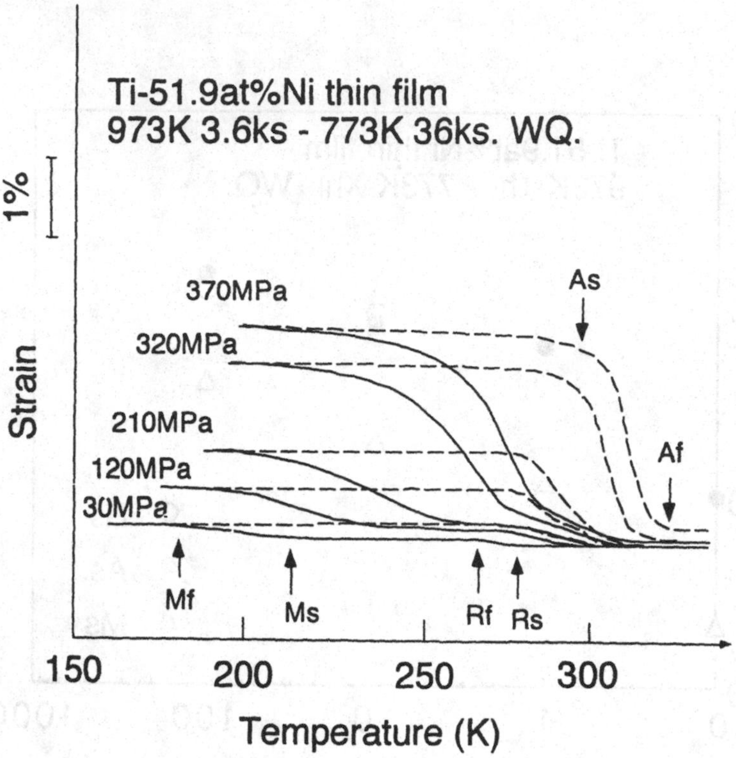

Figure 45 Strain changes due to transformation and reverse transformation upon cooling and heating under various constant stresses in Ti-51.9at%Ni thin film

However, such films will be so brittle that they will fracture by applying stress, hence it is difficult to investigate shape memory characteristics. This may be one of the reasons why many papers on Ti-Ni thin films published in the past did not report the shape memory characteristics.

Figure 45 shows shape memory behavior observed in a Ti-51.9at%Ni thin film which was solution treated at 973K for 3.6ks and followed by aging at 773K for 36ks. The film was cooled and heated under a constant stress. Strain induced by transformations upon cooling is shown by a solid line, while that by a reverse transformation is shown

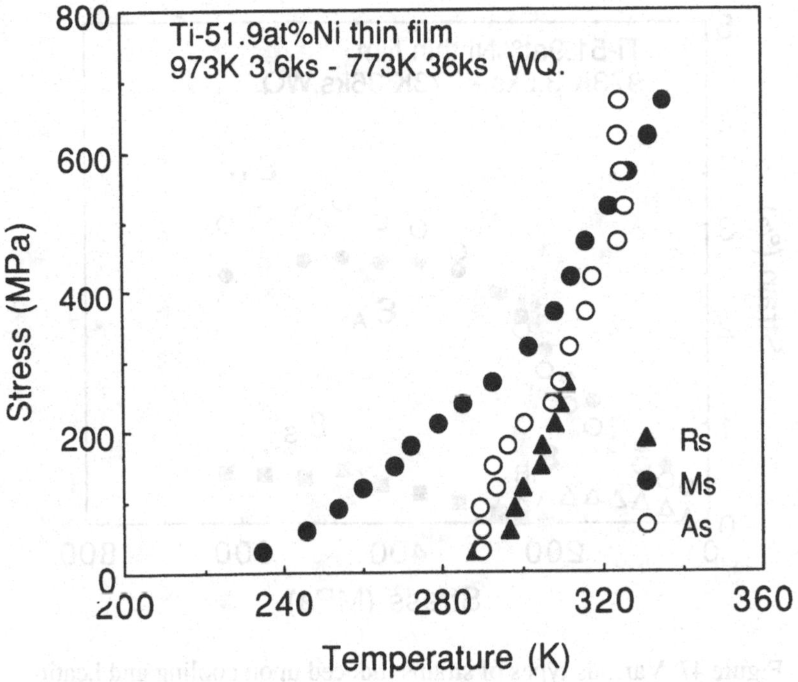

Figure 46 Relationships between stress and transformation temperatures

by a dashed line. Under 30MPa, the first strain change starts to appear upon cooling at *Rs* (268K) which is the R-phase transformation start temperature. Until cooling to *Ms*, the film is in a R-phase. However, the strain keeps increasing with decreasing temperature due to the change in the rhombohedral angle which is shown in Fig. 43. The second strain change starts to appear at *Ms* due to the martensitic transformation, and finishes at *Mf*.

Upon heating, the strain starts to recover at *As* and finishes the shape recovery at *Af* as shown by a dashed line. Since the shape change curves under lower stresses overlap so that there is not enough space to make indications for *As* and *Af*; these temperatures are indicated in the curve obtained under the maximum stress as an example. Transformation temperatures and strain increase with increasing applied stress. This fact implies that the transformation temperatures can also be varied by applying different bias force even in the same film which was subjected to the same heat-treatment.

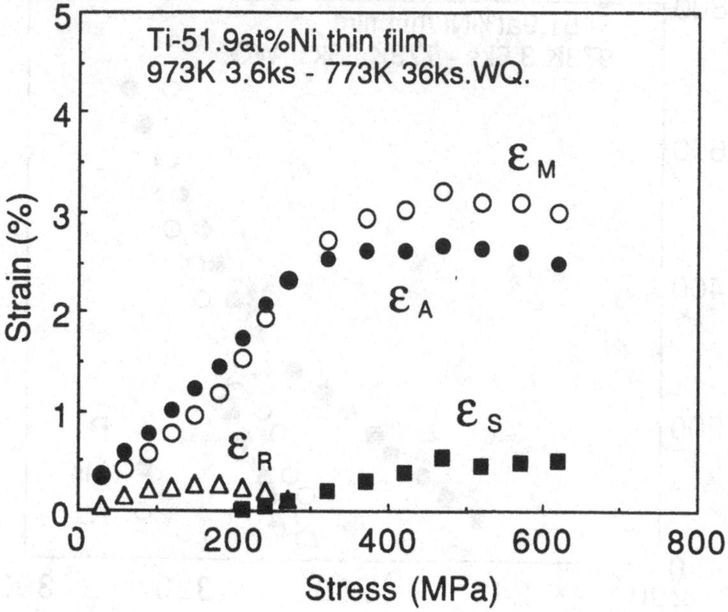

Figure 47 Various types of strains induced upon cooling and heating as a function of stress in Ti-51.9at%Ni thin film

Relationships between the transformation temperatures (*Rs*, *Ms* and *As*) and applied stress are shown in Fig. 46. They show linear relationships. The slope for the reverse martensitic transformation (*As*) is steeper than those for the forward martensitic transformation (*Ms*), causing the temperature hysteresis to become narrower in higher stress region. The slope for the R-phase transformation is steeper than that for the martensitic transformation. The linearity of all these lines indicates that the stress and transformation temperatures satisfy the Clausius-Clapayron relationship.

A two-stage parent(B2)-R-martensite transformation occurs upon cooling and a single stage martensite-B2 reverse transformation occurs upon heating in the thin film as shown in Fig. 45, thus there are three types of transformation strains, i.e., strain(ε_R) due to the B2-R transformation, strain(ε_M) due to the R-martensite transformation and strain(ε_A) due to the martensite-B2 reverse transformation. These strains are plotted as a function of stress in Fig. 47. All of them are of the same order in magnitude for a stress below 100MPa. They increase with increasing stress. However, the increasing rate of ε_R is very small when compared with the other two strains. Besides, the strain ε_R stops increasing above 200MPa, implying that preferentially oriented R-phase variants morphology will be achieved under a low stress. The strains ε_M and ε_A increase rapidly with increasing stress. Therefore, they becomes about ten times larger than ε_R in a higher stress region. The strain ε_A is higher than the strain ε_M by a small amount below a critical stress, where the two lines for the strains intersect each other; because the total strain $\varepsilon_R + \varepsilon_M$ induced upon cooling recovers by a single stage reverse transformation inducing ε_A upon heating. At the higher stress region, the film induce a plastic strain, resulting in suppression of part of the shape recovery upon heating. Therefore, ε_A becomes smaller than ε_M in the higher stress region.

The specimen did not fracture even under 600MPa, although it showed a small plastic strain. Since the plastic deformation is an indication of ductility, the thin film is considered to have a ductile property. Figure 47 shows that the perfect shape memory effect was achieved below about 300MPa, and the recovery stress and strain amount to as much as 600MPa and 3 %, respectively.

3.2 Cu-based Alloys

3.2.1 Phase diagrams

The range of compositions where Cu-Al-Ni shows SME is the region where the single β phase exists at high temperatures, and is limited to around Cu-14Al-4Ni(wt%). The phase diagrams and martensitic transformations of Cu-Al-Ni alloys are basically the same as in Cu-Al binary alloys. Figure 48 shows the phase diagram for binary Cu-Al. There is a β phase region with a BCC structure around the composition of 12wt%Al in the high temperature regions[96] In an equilibrium state, the β phase decomposes into an α phase (FCC) and γ_2 phase (γ brass type structure) at 838K(565C) by the eutectoid decomposition. However, if a specimen is rapidly quenched from the single β phase region, the eutectic decomposition is suppressed and the martensitic transformation occurs at temperatures below the *Ms* temperature. The martensite phase formed in this process differs according to the Al concentration: starting with the low Al concentrations, β', β_1', $\beta_1'+\gamma_1'$, γ_1' and γ_1' phases appear, the subscript indicating a superlattice. The *Ms* temperatures at which these martensite phases are formed can be represented by a single line continuously as shown in the diagram. With Al concentrations more than 11wt%, the disordered β lattice transforms into the ordered β_1 structure (DO_3 or Fe_3Al type structure) at the order-disorder transition temperature *Tc*, which is shown by a single line between the eutectic transformation temperature and *Ms* in the phase diagram. The order-disorder transition cannot be prevented even by rapid quenching, resulting that the martensite phase formed in this composition range takes on an ordered structure by inheriting the order structure of the parent phase.

In the Cu-Al binary system with high Al concentrations, even rapid quenching cannot suppress the precipitation of the γ_2 phase, resulting that thermoelastic martensitic transformations do not occur. Since Ni addition is effective to suppress the diffusion of Cu and Al, the β phase becomes stabler in the ternary Cu-Al-Ni system than in the binary Cu-Al system. Figure 35 shows the cross-section of the phase diagram for Cu-Al-Ni

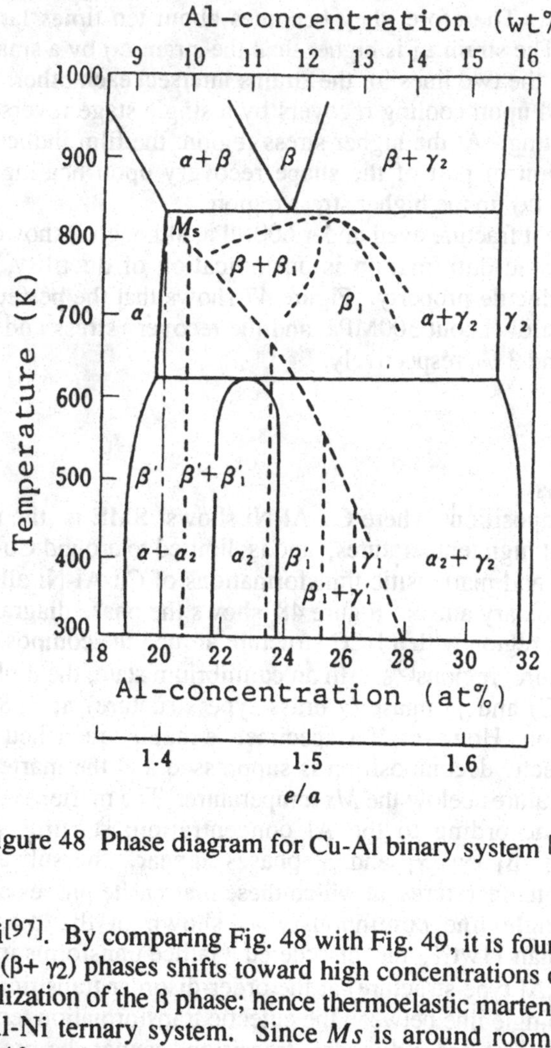

Figure 48 Phase diagram for Cu-Al binary system [96].

with fixed 3wt%Ni[97]. By comparing Fig. 48 with Fig. 49, it is found that the boundary between the β and (β+ γ_2) phases shifts toward high concentrations of Al by adding Ni, indicating the stabilization of the β phase; hence thermoelastic martensitic transformations occur in the Cu-Al-Ni ternary system. Since *Ms* is around room temperature in the vicinity of 14wt%Al as shown in Fig.48, the composition range is used for practical applications, the martensitic phase formed being the γ_1'. In the Al-rich composition range, Ni addition is necessary to suppress the precipitation of the γ_2 phase.

In the Cu-Zn binary system as shown in Fig.50 [98], *Ms* is too low in the composition range where thermoelastic transformations appear. Therefor, it is necessary to adjust the transformation temperature by adding a third element such as Al, Ge, Si, Sn or Be. Figure 51 shows the cross-section of the phase diagram of Cu-Zn-Al ternary system with fixed 6wt%Al[99]. By comparing Fig.51 with Fig.50, it is found that

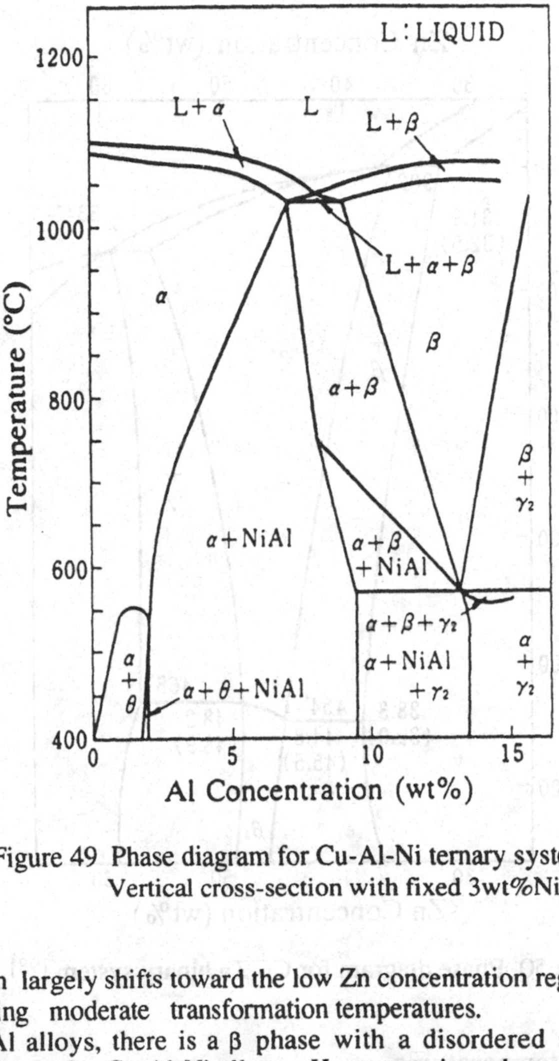

Figure 49 Phase diagram for Cu-Al-Ni ternary system.
Vertical cross-section with fixed 3wt%Ni [97].

the β phase region largely shifts toward the low Zn concentration region by adding Al, resulting achieving moderate transformation temperatures.

In Cu-Zn-Al alloys, there is a β phase with a disordered structure in a high temperature region as for Cu-Al-Ni alloys. However, since there is an order-disorder transition at an intermediate temperature, the β2 ordered structure (B2 or CsCl type) is formed during quenching; this is the parent phase of the Cu-Zn-Al alloys. In the range of Al-rich compositions, though, a B2-DO_3 ordering transition occurs at a relatively high temperature region. Therefore, the DO_3 structure appears in Cu-Zn-Al with the Al-rich composition.

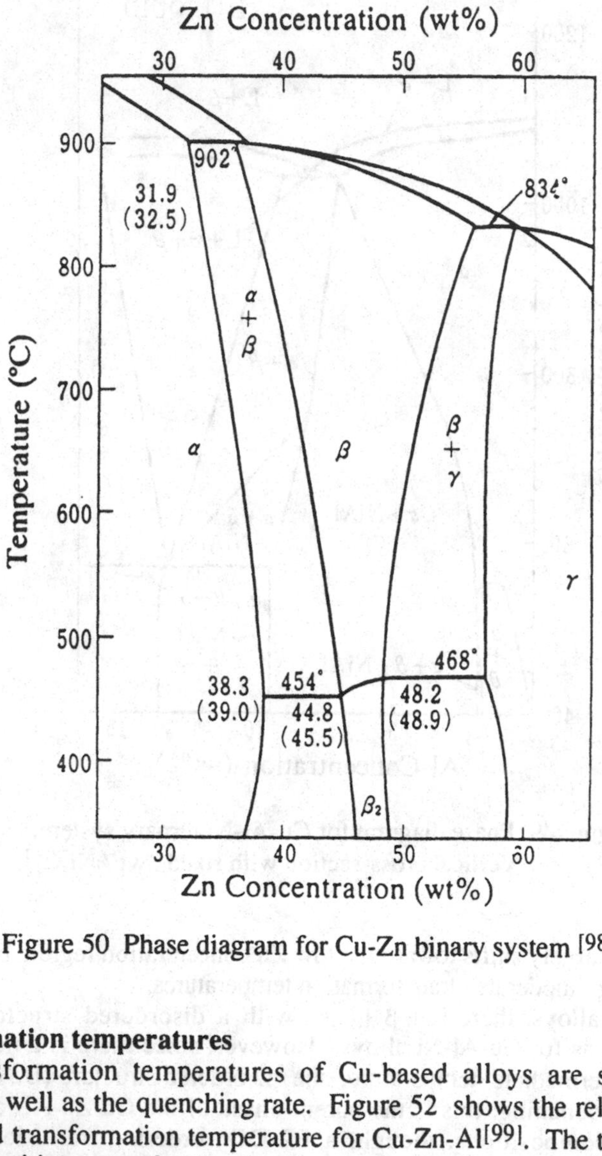

Figure 50 Phase diagram for Cu-Zn binary system [98].

3.2.2 Transfomation temperatures

The transformation temperatures of Cu-based alloys are sensitive to their compositions as well as the quenching rate. Figure 52 shows the relation between the composition and transformation temperature for Cu-Zn-Al[99]. The triangle on the left shows the composition range for the ternary system; the black colored area is magnified to produce the right figure, where the measured *Ms* temperature are shown.

Figure 53 shows the transformation temperatures vs. Al concentration relationship in Cu-Al-Ni alloys with the fixed Ni concentration at 4.0at% [100]. The straight lines represent the averages for more than 50 single crystals and polycrystals, the scatter

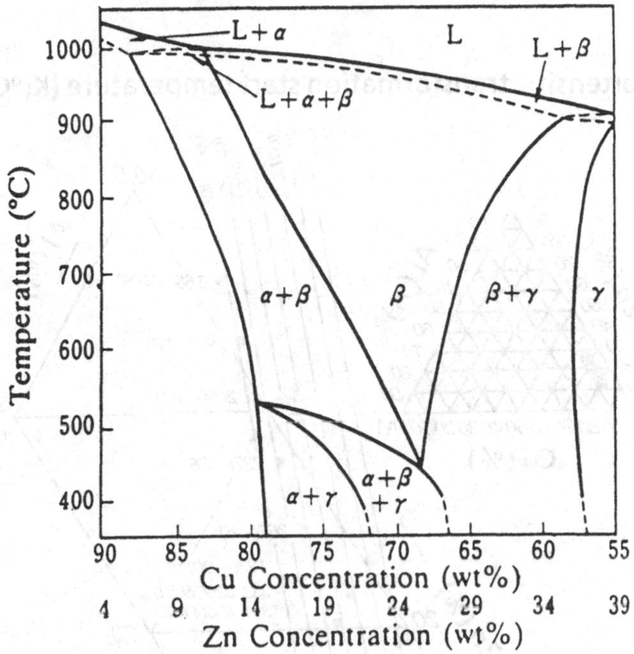

Figure 51 Phase diagram for Cu-Zn-Al ternary system fixed 6wt%Al [99].

being about 20K to 30K. The transformation temperature decrease with increasing Al concentration. Even though the single crystal and polycrystal were produced from the same ingot, the *Ms* of the single crystal was about 20K higher than that of the polycrystal. The lower *Ms* of the polycrystal is due to the constraining forces from the surronding grains[101,102].

The Ni concentration also changes the transformation temperatures as shown in Table 4, although the effect is not so large as the Al concentration [103]. The transformation temperatures decrease with increasing Ni concentration, the Al concentration being fixed. Finally, the coolig rate also affects the transformation temperatures, as shown in Table 5[104]. The effect of the cooling rate may be caused by variations in the Al concentration in the matrix due to the precipitation of the γ_2 phase, exess vacancies frozen in by quenching, variations in the degree of order, or thermal stresses induced during quenching. However, the mechanism for the shift of the transformation temperature is not clearly understood at present.

Thermal cycling also changes the transformation temperatures due to introducing dislocations or changing the degree of order [109-113].

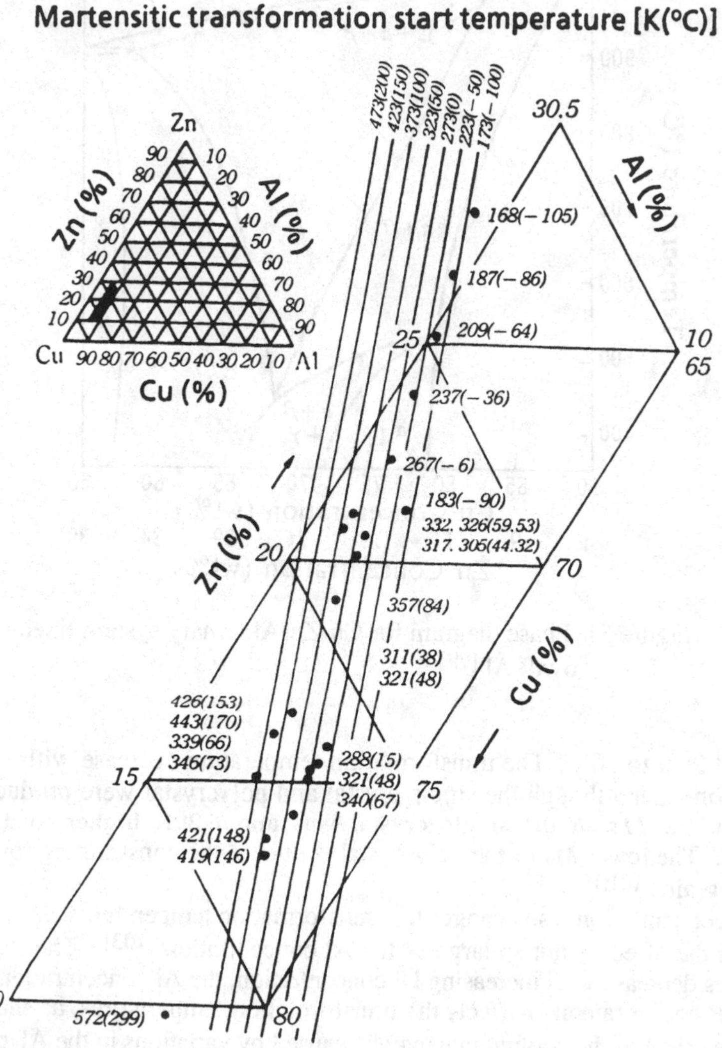

Figure 52 Relation between composition and M_S in Cu-Zn-Al alloy [99].

Table 4 Relation between Ni-content and transformation temperatures in Cu-Al-Ni alloys All specimens were water-quenched after solid solution-treatment at 1193K[103].

Cu (wt%)	Al (wt%)	Ni (wt%)	Ms (K)	Mf (K)	As (K)	Af (K)
72	28	0	285	240	230	320
71	28	1	265	250	277	290
70	28	2	270	269	283	290
69	28	3	248	233	260	280
68	28	4	180	172	200	230

Table 5 Relation between cooling rate and M_S in Cu-Al-Ni alloy. All specimens were quenched from 1273 K into 288 K or 373 K. The cooling rate was changed by varying the temperature of the quenching media(T_{gm}) [104].

Composition (wt%)	Tqm (K)	Ms (K)	Tqm (K)	Ms (K)
Cu-14.0Al-3.9Ni		282		363
Cu-14.1Al-4.0Ni	288	262	373	333
Cu-14.2Al-4.0Ni		228		308

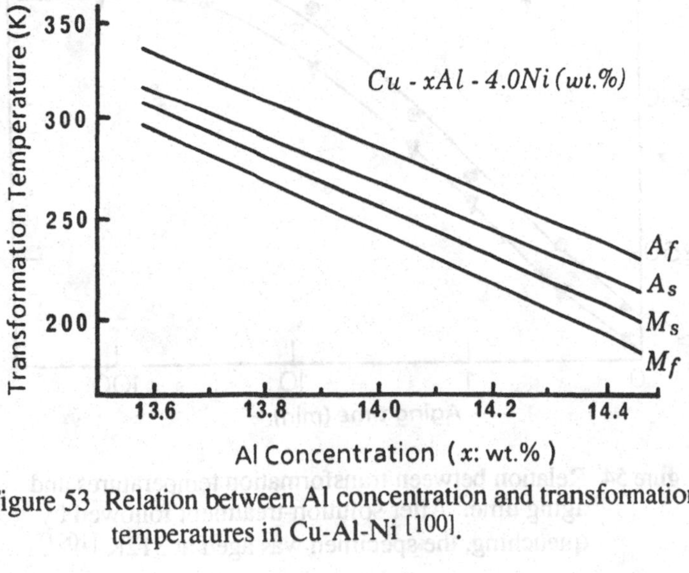

Figure 53 Relation between Al concentration and transformation temperatures in Cu-Al-Ni [100].

3.2.3 Aging effect

There are several causes conceivable for the effects of aging. The aging effects appear differently according to the aging temperature and depending upon whether the aging occurs in the parent phase or in the martensitic phase. In any case, the aging effects are associated with atomic diffusion. Thus the effects of the addition of elements which change the diffusion coefficients are also important. These problems will be described in the following.

(a) Aging in the parent phase

Generally, the following two factors are considered causes of aging effects in Cu-based alloys : (1) change of the degree of order after quenching and (2) the formation of precipitates.

In A_3B type alloys, The ordering reaction during quenching occurs in two stages; i.e. a disordered lattice changes into an B2 superlattice at higher temperatures, and then the latter into a DO_3 superlattice. Since the β-B2 ordering transition occurs at a temperature around 773K(500°C) in the Cu-Zn-Al system, this process will be completed after quenching. The B2-DO_3 transition shows a strong composition dependence and can be lowered even to room temperature. In such a case, an incompletely ordered structure may be obtained by quenching.

If specimens thus obtained are aged at 312K, the temperatures M^* and A^*, which are the temperatures measured by a differential scanning calorimeter and associated with *Ms*

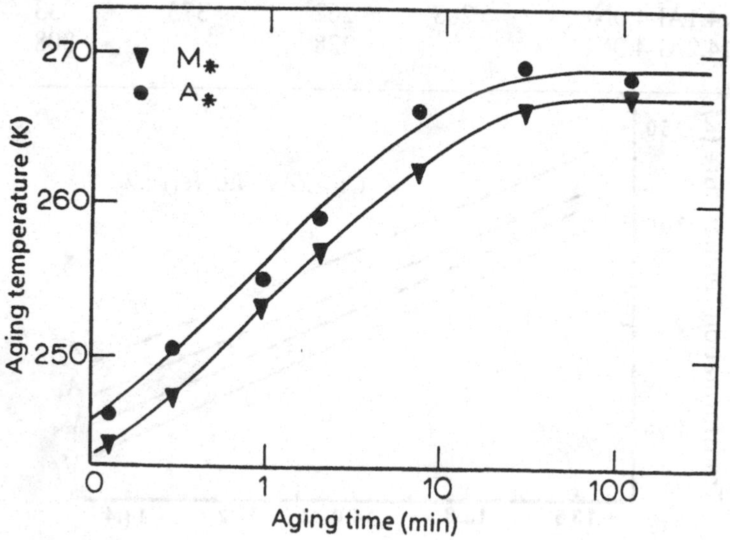

Figure 54 Relation between transformation temperatures and aging time. After solution-treatment followed by quenching, the specimen was aged at 312K [109].

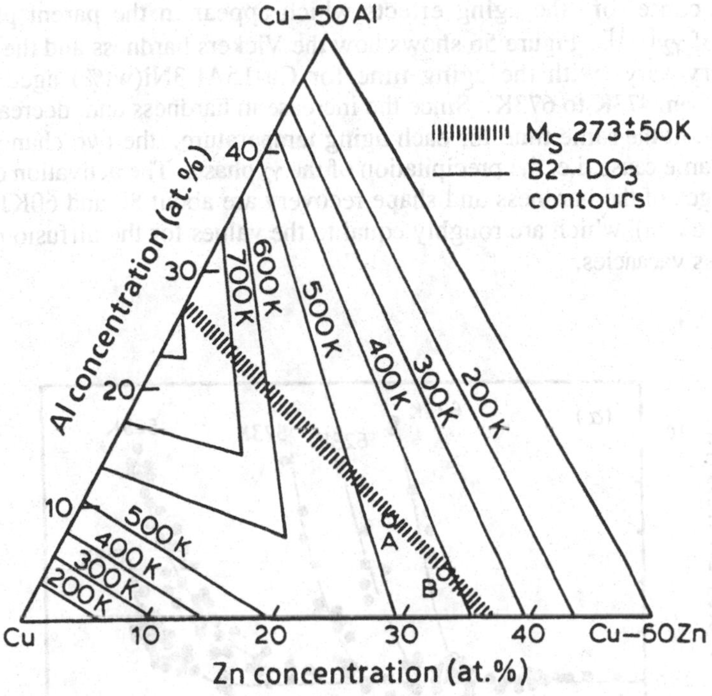

Figure 55 Calculated results of the relation between the composition and the B2→DO3 order-disorder transition temperature in Cu-Zn-Al [109].

and *As* respectively, increase with increasing aging time as shown in Fig.54 [109]. After 20~30 minutes, both *M** and *A** increase about 20K and then become constant. If these specimens were subjected to flash annealing at temperatures above the B2-DO3 transition temperature followed by quenching, *M** and *A** reverted to the values before aging.

Since the B2-DO3 ordering transition temperature depends strongly upon the alloys composition[110], the composition affects the structure obtained by quenching. Figure 55 shows the calculated results of the B2-DO3 ordering transition temperature in the Cu-Zn-Al ternary system [109]. The shaded area in the figure shows the range of composition for which *Ms* is about 323K(50°C), indicating that the transition temperature increases with increasing the Al concentration in the shaded area keeping the *Ms* unchanged. When the ordering transition occurs at higher temperatures, the B2-DO3 atomic rearrangement occurs easily to form a complete DO3 structure. Therefore, Cu-Zn-Al alloys with high concentrations of Al will be of the DO3 structure, and those with low Al concentrations will have the B2 structure.

Another cause of the aging effects which appear in the parent phase is the precipitation of γ_2 [111]. Figure 56 shows how the Vickers hardness and the percentage shape recovery vary with the aging time for Cu-15Al-3Ni(wt%) aged at various temperatures from 473K to 673K. Since the increase in hardness and decrease in shape recovery begin at the same time for each aging temperature, the two changes must be based on the same cause, i.e. the precipitation of the γ_2 phase. The activation energies for both the changes of the hardness and shape recovery are about 80 and 60KJ/mol(about 0.83 and 0.62 eV/at), which are roughly equal to the values for the diffusion associated with the excess vacancies.

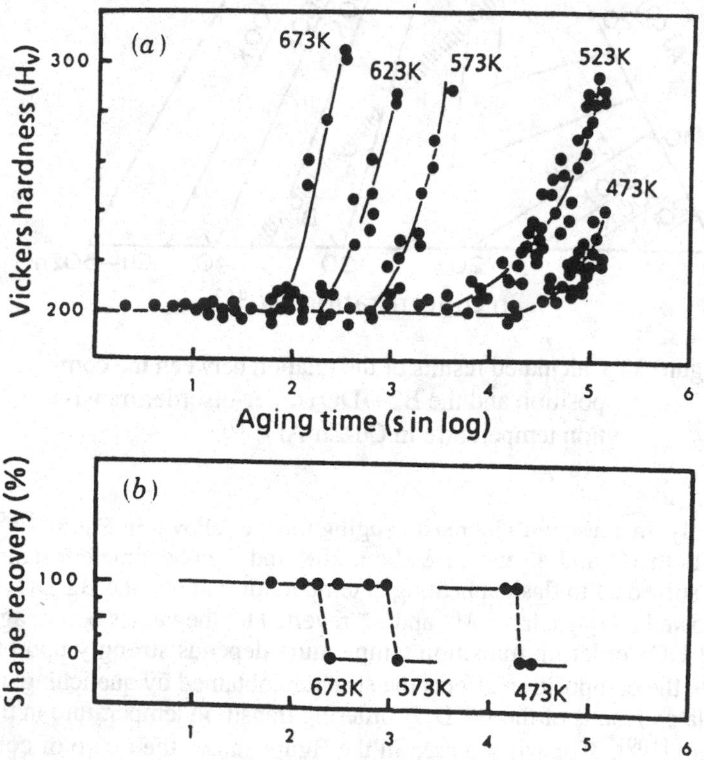

Figure 56 (a) Change in the Vickers hardness and (b) the shape recovery ratio as a function of aging in Cu-Al-Ni [111].

(b) Aging in the martensitic phase

Generally, the martensitic transformations occur by the cooperative movement of atoms, and thus the martensite inherits the ordered atomic arrangement of the parent phase. Accordingly, a possibility remains of further reducing the free energy by the rearrangement of the atoms in the martensitic phase [97,109,112]. This phenomenon brings about the increase of the reverse-transformation temperature, because the martensitic phase becomes more stable. If a SMA is used as a temperature detector and an actuator utilizing the force produced by the reverse-transformation, the aging in the martensitic phase will change the operation temperature after the SMA is installed. Since the rearrangement of atoms due to the aging in the martensitic phase is also associated with diffusion, it is important to suppress diffusion in order to stabilize the SME characteristics in practical applications.

(c) Effect of the third element on aging

The hardness of Cu-Al-Ni alloys with different Ni concentrations was measured as a function of aging temperature as shown in Fig. 57 for a fixed aging time of 10min [103]. The point at which the hardness begins to rise corresponds to the temperature where γ_2 precipitation begins. The point shifts to higher temperatures when the Ni concentration was increased. This indicates that the Ni operates to suppress the diffusion of the Cu and Al. In fact, in binary Cu-Al alloys, diffusion is rapid and the precipitation of the γ_2 cannot be suppressed even by quenching.

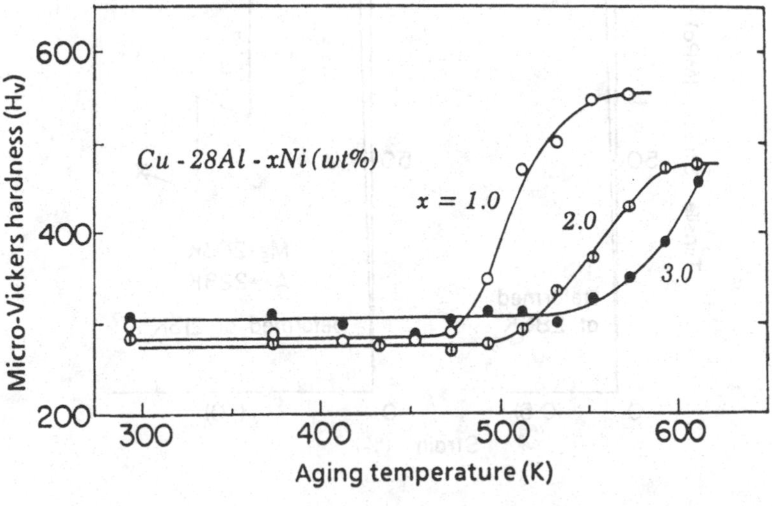

The aging time: 10min
x: Ni concentration

Figure 57 Relation between aging temperature and micro-Vickers hardness in Cu-Al-Ni alloys. [103].

3.2.4 Fracture and Fatigue

One of the most serious problems in applying Cu-based alloys are the brittleness due to intergranular fractures[104,113,114]. The following are considered possible causes of the intergranular fractures:

(1) the large elastic anisotropy,
(2) large grain size,
(3) the large orientation dependence of the transformation strain, and
(4) grain boundary segregation.

Cases (1)~(3) may produce intergranular fracture through large stress concentration along grain boundaries. Cause (4) lowers the strength of the boundaries themselves and results in intergranular fractures.

In order to clarify the cause for the intergranular fracture, well designed bicrystals are ideal specimens[115,116]. Six bicrystals were used for the purpose. They can be classified into three types according to the relative orientations of the two component crystals as follows:

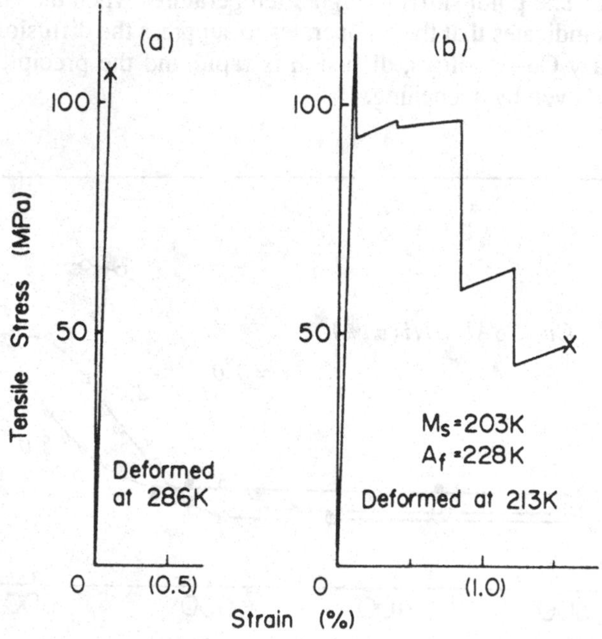

x: Final fracture point

Figure 58 Stress-strain curves for asymmetric bicrystal 2, the deformation temperature being (a) 286K and (b) 213K [115].

(1) Asymmetric bicrystals 1, 2 and 3 with stress concentrations at the grain boundaries due to anisotropic elasticity,
(2) Asymmetric bicrystal 4 which have a special orientation relationship which does not produce stress concentrations upon elastic deformation, but produce upon transformation, and
(3) a symmetric bicrystal which does not produce any stress concentration in the boundary either elastically or after the transformation.

Asymmetric bicrystal 1 was a combination of component crystals in whith the difference between elastic strains at the grain boundary became extremely large. The intergranular fracture occurred in this specimen due to thermal stresses when it was quenched. The difference in the elastic strains at the grain boundary in asymmetric bicrystals 2 and 3 was smaller than in asymmetric bicrystal 1, and the specimens did not crack during quenching. However, bicrystal 2 fractured in the elastic deformation stage as shown in Fig.58. If we assume the cause of the cracking in this case was stress concentration due to the difference in transformation strain in the grain boundary, the fracture stress should show a temperature dependence as the stress for inducing martensite does. However, grain boundary cracking occurred almost at the same stress irrespective of deformation temperature. Therefore, it can be concluded that the cause of the intergranular fracture in asymmetric bicrystal 2 was the stress concentrations due to the large elastic anisotropy of this alloy.

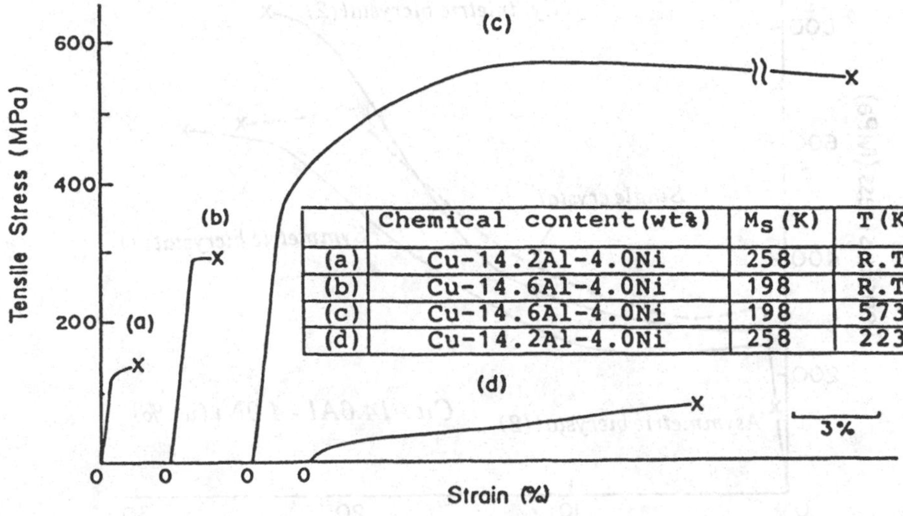

	Chemical content (wt%)	M_S (K)	T (K)
(a)	Cu-14.2Al-4.0Ni	258	R.T.
(b)	Cu-14.6Al-4.0Ni	198	R.T.
(c)	Cu-14.6Al-4.0Ni	198	573
(d)	Cu-14.2Al-4.0Ni	258	223

Figure 59 Effect of composition and test temperature on the stress-strain curves in asymmetric bicrystal 4 [116].

In the special asymmetric bicrystal 4, stress concentrations do not arise elastically but are instead generated on the grain boundary by the difference in transformation strains in the component crystals. This specimen did not fracture elastically eventhough the deformation temperature or the composition were altered, however, intergranular fractures always occurred after the stress induced martensitic transformation as shown in Fig. 59. Therefore, the cause for the intergranular fracture lies in the stress concentrations due to the large difference in the transformation strains of the component crystals and hence the fracture stress depends on the deformation temperature or the composition in this case.

In the symmetric bicrystal, both component crystals rotated around [100] and the tensile axis inclined mutually and symmetrically 10 degrees away from the [001] direction in each crystal. Fracture occurred transgranularly after deformations more than 20% as shown in Fig.60, where the stress-strain curves for single crystal, the symmetric bicrystal and asymmetric bicrystal 2 are compared.

An auger electron analysis revealed that oxygen atoms were detected at grain boundaries in a Cu-Al-Ni alloy, while they are not in a Ti-doped Cu-Al-Ni alloy; the Cu-Al-Ni alloy showed intergranular fracture, while the Cu-Al-Ni-Ti alloy transgranular fracture [127,128]. This fact suggests that the segregation of oxygen at grain boundaries is also one of the causes of intergranular fracture.

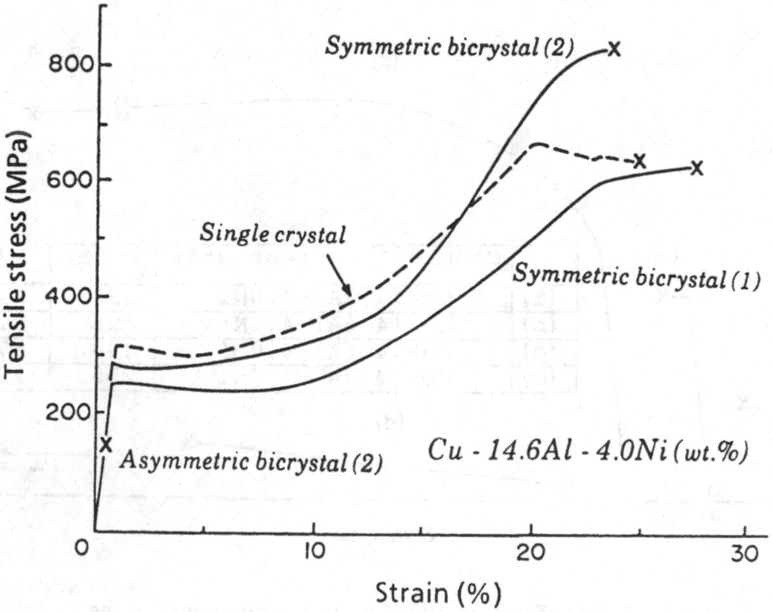

Figure 60 Stress-strain curves for a single crystal and symmetric bicrystal of Cu-Al-Ni [116]. x: Fracture point

Table 6 Relation between fracture patterns and various factors[12].

Alloy	Anisotropic factor	Dependence of transformation strain on orientation	Parent phase	Slip stress (MPa)	Fracture pattern
CuAlNi	13	Large	DO_3	~600	Intergranylar fracture
CuZnAl	15	Large	B2	~200	Transgranular fracture
		Large	DO_3	High	Intergranylar fracture
TiNi	2	Large	B2	~100	Transgranular fracture

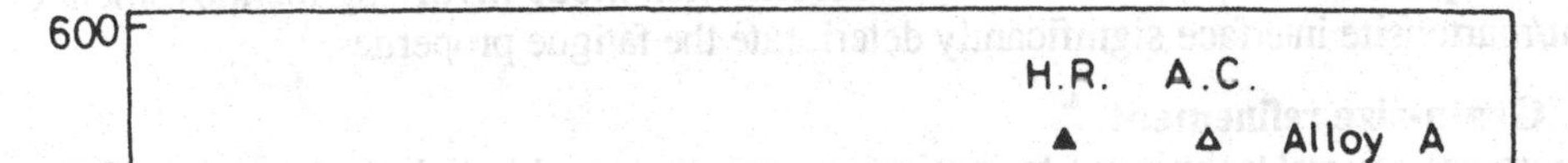

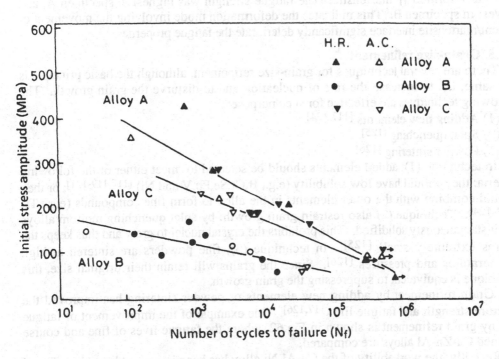

Figure 61 Fatigue life of single crystals of Cu-Al-Ni [114].

Table 6 shows the observed fracture pattern of Cu-Al-Ni, Cu-Zn-Al and Ti-Ni alloys along with the elastic anisotropy factor, orientation dependence of transformation strain, the crystal structure of the parent phase, stress for slip. Although large stress concentrations due to the strong orientation dependence of transformation strain can be expected in every alloy listed in the table, Cu-Zn-Al with B2 structure and Ti-Ni shows transgranular fracture. Since the critical stress for slip is low(100~200MPa) for both cases, it seems that the stress concentration can be effectively relaxed by slip deformation. On the other hand, since Cu-Al-Ni and Cu-Zn-Al with the DO_3 structure have high stress for slip, it is not effective for these alloys to relax the large stress concentration; and hence grain boundary fracture occurs. Therefore, it is concluded that the critical stress for slip is also one of the important factors controlling the fracture pattern.

The fatigue life depends strongly upon the cyclic deformation mode [114]. Fig.61 shows the fatigue life of polycrystalline Cu-Al-Ni specimens(A,B and C) with different concentrations. The *Ms* temperature for specimen A is 180K, and specimen B was 273K, and specimen C was 420K. Consequently, the deformation mode at room temperature for specimen A is elastic deformation in the parent phase, for specimen B it is the stress-induced transformation, and for specimen C it was the rearrangement of thermally formed γ_1' martensite. The fatigue strength was highest in specimen A, and lowest in specimen B. This indicates the deformation mode involving the movement of parent/martensite interface significantly deteriorate the fatigue properties.

3.2.5 Grain-size refinement

There are several techniques for grain-size refinement, although the basic principle is the same, i.e. to increase the rate of nucleation and to disturve the grain growth. The following techniques are effective for such purpose:

(1) Adding new elements [117-124]
(2) Splat quenching [125]
(3) Powder sintering [126]

In technique (1), added elements should be selected to meet either of the following criteria: they should have low solubility (e.g., B,Cr,Se,Pb,V and Ni) [117,119-121], or they should combine with the other elements in the alloy to form fine compounds (e.g., Ti) [122-124]. Technique (2) also restrain grain growth: by splat quenching moltom alloys are instantaneously solidified. This prohibits the crystal nuclei to grow and thus keeps the grains extremely small [125]. In technique (3), fine powders are sintered at high temperatures and pressures [126]. Since the grains will retain their original size, this technique is equivalent to suppressing the grain growth.

Grain refinement by adding new elements or powder sintering has improved the fracture strength and fatigue life [117,126]. One example of the improvement of fatigue life by grain refinement is shown in Fig.62, where the fatigue lives of fine and course grained Cu-Zn-Al alloys are compared.

Recently, the workability of the Cu-Al-Ni alloy has been improved by adding Ti and Mn[123,127]. The concept of the alloy designing is explained in the following. Grain refinement was achieved by adding Ti, resulting in a good hot workability. In order to improve the cold-workability, Al concentration was reduced from 14at% (hyper-

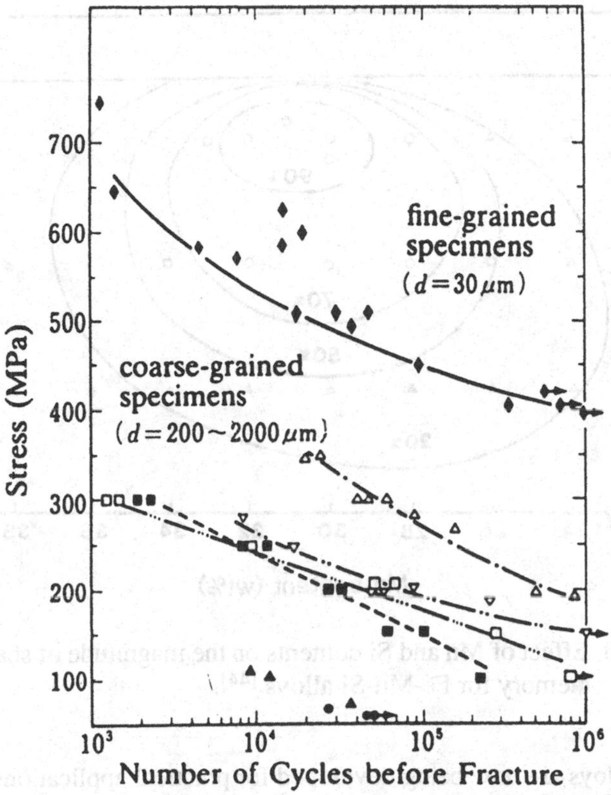

Figure 62 Fatigue life in fine- and coarse-grained specimens [126].

eutectoid composition) to 12at% (hypoeutectoid composition). In a hypereutectoid composition region, the γ_2 precipitates are produced, deteriorating the cold-workability. However, in a hypoeutectoid compositions region, there are two phases, i.e. α and β, above the eutectoid transformation temperature. Since the α phase is FCC structure, it is easily deformed, resulting in a good cold-workability (~10% at room temperature). However, it is further necessary to decrease the *Ms* temperature which has increased by reducing the Al concentration (e.g. 1at% increment of Al causes the *Ms* to increase about 120K). For that purpose, the addition of Mn is effective, the final product being Cu-12%Al-5%Ni-2%Mn-1%Ti(at).

3.3 Fe Alloys

Among ferrous SMAs in Table 2, Fe-Pt [128-131] and Fe-Pd [132] have been investigated only for the academic interest, because Pt and Pd are very expensive elements.

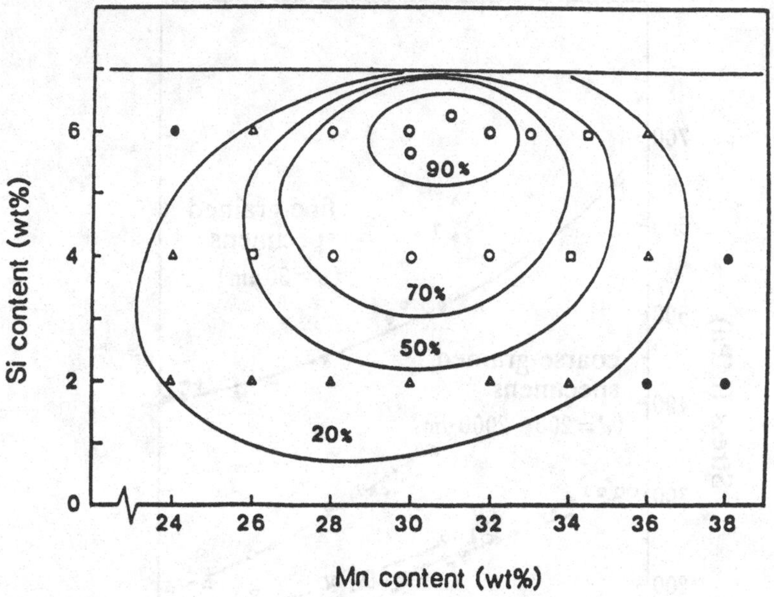

Figure 63 Effect of Mn and Si contents on the magnitude of shape memory for Fe-Mn-Si alloys [44].

The other ferrous alloys are now being developed for practical applications. One of the common features of the ferrous SMAs is that the martensite morphology is thin plate type [43].

The thin plate martensite is characterized by the following features; i.e. the lattice invariant shear is perfectly twinning and the stress due to the transformation strain is accommodated by the elastic deformation instead of the plastic deformation. Hence, dislocations are rare in the austenite in the vicituty of the thin plate martensite, resulting in the formation of the mobile parent/martensite interface. In order to achieve the above situation, the following four factors are considered to be effective[133].

(1) The yield stress of the parent phase is high and/or the elastic modulus is low.
(2) The volume change and transformation shear strain upon transformation are small.
(3) The tetragonality (c/a) of the martensite is large. The twinning shear and the twinning boundary energy decrease with increasing c/a; hence, the twinning becomes easy in the martensite.
(4) *Ms* is low. The twinning becomes easier to occur compared to the slip deformation with decreasing *Ms*; The yield stress of the parent phase also increases.

All the above four factors make slip deformation difficult to occur. In cases of Fe-Ni-C [134] and Fe-Al-C [135] alloys, the transition temperature for martensite morphology from lenticular to thin plate increases with increasing C-content, indicating that the thin plate martensite becomes easier to appear with increasing C-content. Besides, the tetragonality of the martensite and the yield stress of the parent phase increase with increasing C-content. High C-content also makes the Curie temperature (*Tc*) increase, resulting in a large difference between *Tc* and *Ms* and hence a small volume change upon transformation due to the invar effect. The above characteristics of Fe-Ni-C and Fe-Al-C alloys are effective to form the thin plate martensite and hence SME.

However, these alloys include carbon atoms which combine with Ni or Al to form carbides in the martensite upon reverse-transformation, causing the SME to occur incompletely by suppressing the mobility of the parent/martensite interface. Therefore, several new ferrous alloys have been developed, i.e. Fe-Ni-Ti-Co, Fe-Mn-Si and Fe-Cr-Ni-Mn-Si-Co alloys.

The characteristics of each of these alloys are explained briefly in the following.

(1) Fe-Ni-Co-Ti [43] :

The martensitic transformation of the Fe-33%Ni-4%Ti-10%Co(wt%) alloy becomes thermoelastic with a small thermal hysteresis if it is ausaged at 793K for 1.8Ksec. The martensitic transformation occurs from a FCC austenite (γ) to BCT (α'). The effect of each element is as follows: High Ni-content makes *Ms* decrease. Addition of Ti causes ordered fine γ' (Ni_3Ti) precipitates to be formed, resulting in a high yield stress and a large back stress for assisting the reverse-transformation. Addition of Co makes the Qurie temperature increase, resulting in a small volume change upon transformation and a low modulus. One difficulty in practical application of this alloy is at present the low value of *Ms* (below 200 K).

(2) Fe-Mn-Si [44,45] :

The SME is associated with the stress-induced transformation from the FCC austenite to ε(HCP) martensite. If the specimen is cooled below *Ms*, thermally induced martensites are formed, suppressing part of the shape recovery. Mn is effective to reduce the stacking fault energy of the austenite and hence the ε phase is easily stress-induced. Since Mn reduces *Ms*, the yield stress of the austenite increases. However, Mn raises the Neel temperature (T_N). If T_N becomes higher than *Ms*, the stress-induced martensitic transformation is suppressed. However, Si is effective to decrease T_N. Si is also effective to reduce the stacking fault energy and raise the yield stress of the austenite; assisting a good SME. Therefore, there is an optimum range for Si- and Mn-contents in order to achieve good SME. Figure 63 shows the effect of Mn- and Si-contents on the magnitude of shape recovery. The composition range of Fe-28~33%Mn-5~6%Si shows nearly perfect shape recovery. One of the advantages of this alloy is that *Ms* is around room temperature.

(3) Fe-Cr-Ni-Mn-Si-Co [136]:

The SME is associated with the γ-ε' stress-induced transformation, similar to the Fe-Mn-Si alloy. It was recently reported that the composition range of 7~15%Cr, <10%Ni, <15%Mn, <7%Si and 0~15%Co shows complete shape recovery if the deformation does not exceed 4% strain. Another advantage of this alloy is a high corrosion-resistance as a stainless steel. This alloy also has an advantage in *Ms*, because the *Ms* is between 173 K and 323 K.

4. PROSPECTS IN APPLICATIONS OF SHAPE MEMORY ALLOYS

As mentioned in the introduction, the total number of patents for applications of SMAs amounts to more than 15,000 [41]. They include almost all kinds of industrial fields, i.e. electrical engineering, machinery, transportation, chemical engineering, energy, medicine, and so on. This fact indicates a great demand for applications of SMAs. However, the number of application products in the market now amounts to only 2 or 3 % of the total number of the applied patents. One of possible reasons for this is probably the current price of Ti-Ni alloys.

However, most of the current price of Ti-Ni alloys is not the cost of raw materials but the manufacturing cost. Thus, the price of Ti-Ni alloys is expected to further decrease in the near future if the amount of alloy production increases. Besides, if the characteristics of Cu-based and ferrous alloys are further improved to the level suitable for applications, there will be no problem of cost.

SMA and PE characteristics of Ti-Ni alloys have been well developed by using special thermo-mechanical treatments [13,137], and they are now in use in actual applications. Examples of such applications include pipe joints, rimless glasses, brassieres, PE arch wires for orthodontics etc. which require simple functions. However, there are many other applications which demand more precise and/or repeating functions. In order to apply Ti-Ni alloys for such purposes, further improvement is still necessary. This is another reason for the small number of products in the market at present. Some of the key directions for the further improvement will be the addition of the third elements, alloy purification, developing a special texture, and so on.

Among several types of high performance materials proposed for fabricating microactuators, Ti-Ni shape memory alloy thin films exhibit significant advantages including large deformation and recovery force. The disadvantage of the bulk shape memory alloys, a slow response due to the limitation of cooling rate, can be greatly improved in micro size actuators. Recent success in producing Ti-Ni thin films by sputtering will open a door to a new application field [95,138], where variety of microactuators made of Ti-Ni thin films will power small machines and robots of micron sizes.

REFERENCES

1) L. C. Chang and T. A. Read: Trans. AIME, 189 (1951), 47.
2) M. W. Burkart and T. A. Read: Trans. AIME, 197 (1953), 1516.
3) Z. S. Basinski and J. W. Christian: Acta Met., 2 (1954), 101.
4) W. J. Buehler, J. V. Gilfrich and K. C. Weiley: J. Appl. Phys., 34 (1963), 1467.
5) I. A. Arbuzova and L. G. Khandros: Fiz. Met. Metalloved, 17 (1964), 390.
6) K. Otsuka and K. Shimizu: Scripta Met., 4 (1970), 469.
7) K. Otsuka, Jap. J. Appl. Phys., 10 (1971), 571.
8) T. Saburi and S. Nenno: Proc. Int. Conf. Solid-solid Phase Transformations, Pittsuburgh,
PA, (1981), 1455.
9) T. Saburi, C. M. Wayman, K. Takata and S. Nenno: Acta Metall., 28 (1980), 15.
10) K. Otsuka, C.M. Wayman, K. Nakai, H. Sakamoto and K. Shimizu: Acta Metall., 24 (1976), 207.
11) K. Otsuka, H. Sakamoto and K. Shimizu: Acta Metall., 27 (1979), 585.
12) S. Miyazaki and K. Otsuka: Shape Memory Alloys, ed. H. Funakubo, Gordon and Breach Science Publishers, (1987), 116.
13) S. Miyazaki, Y. Ohmi, K. Otsuka and Y. Suzuki: J. de Phys., 43, Suppl. 12, (1982), C4-255.
14) T. Saburi, T. Tatsumi and S. Nenno: J. de Phys., 43, Suppl. 12, (1982), C4-261.
15) S. Miyazaki, K. Otsuka and Y. Suzuki: Scripta Metall., 15 (1981), 287.
16) S. Miyazaki, T. Imai, K. Otsuka and Y. Suzuki: Scripta Metall., 15 (1981), 853.
17) F. Takei, T. Miura, S. Miyazaki, S. Kimura, K. Otsuka and Y. Suzuki: Scripta Metall., 17 (1983), 987.
18) S. Miyazaki, S. Kimura, F. Takei, T. Miura, K. Otsuka and Y.Suzuki: Scripta Metall., 17 (1983),1057.
19) S. Miyazaki, T. Imai, Y. Igo and K. Otsuka: Met. Trans. A, 17 (1986), 115.
20) T. Honma: Proc. ICOMAT-86,Nara, (1986), 709.
21) S. Miyazaki and C. M. Wayman: Acta Metall., 36 (1988), 181.
22) S. Miyazaki, K. Otsuka and C. M. Wayman: Proc. MRS Inter. Meeting on Advanced Materials, Tokyo (1988), 93.
23) M. Matsumoto and T. Honma: Proc. 1st JIM Int. Symp. on New Aspects of Martensitic Transformation, Suppl. to Trans. JIM, 17 (1976), 187.
24) C.M. Hwang, M. Meichle, M. B. Salamon and C.M. Wayman: Phil. Mag. A; 47 (1983), 31.
25) M.B. Salamon, M.E. Meichle and C.M. Wayman: Phys. Rev. B, 31 (1985), 7306.
26) V. N. Khachin, V. E. Gunter, V. P. Sivokha and A. S. Savinov: Proc. ICOMAT-79, Boston, (1979), 474.
27) H. C. Ling and R. Kaplow: Met. Trans. A, 11 (1980), 77.
28) H. C. Ling and R. Kaplow: Met. Trans. A, 12 (1981), 2101.
29) S. Miyazaki and K. Otsuka: Phil. Mag. A, 50 (1984), 393.
30) S. Miyazaki and K. Otsuka: Met. Trans. A, 17 (1986), 53.
31) S. Miyazaki, S. Kimura and K. Otsuka: Phil. Mag. A, 57 (1988), 467.

32) K. M. Knowles and D. A. Smith: Acta Metall., 29 (1981), 101.
33) T. Saburi, M. Yoshida and S. Nenno: Scripta Metall., 18 (1984), 363.
34) S. Miyazaki, S. Kimura and K. Otsuka: Scripta Metall., 18 (1984), 883.
35) Y. Kudoh, M. Tokonami, S. Miyazaki and K. Otsuka: Acta Metall., 33 (1985), 2049.
36) O. Matsumoto, S. Miyazaki, K. Otsuka and H. Tamura: Acta Metall., 35 (1987), 2137.
37) M. Nishida, C. M. Wayman and T. Honma: Met. Trans. A, 17 (1986), 1505.
38) T. Tadaki, Y. Nakata, K. Shimizu and K. Otsuka: Trans. JIM, 27 (1986), 731.
39) M. Nishida, C. M. Wayman, R. Kainuma and T. Honma: Scripta Metall., 20 (1986), 899.
40) T. Saburi, S. Nenno and T. Fukuda: J. Less-Common Metals, 125 (1986), 157.
41) Report on the Patents for Shape Memory Alloys, ed. Association of Shape Memory Alloys, (1984-94).
42) Y. N. Koval, V. V. Kokorin and L. G. Khandros: Phys. Met. Metall., 48 (1981), 162.
43) T. Maki and I. Tamura: Proc. ICOMAT-86, Nara, (1986), 963.
44) M. Murakami, H. Otsuka, H. G. Suzuki and S. Matsuda: Proc. ICOMAT-86, Nara, (1986), 985.
45) A. Sato, E. Chishima, K. Soma and T. Mori: Acta Metall., 30 (1982), 1177.
46) Z. Nishiyama: Martensitic Transformations, New York Academic Press, (1978).
47) H. Warlimont and L. Delaey: Prog. Mater. Sci., 18 (1974).
48) N. Nakanishi: Prog. Mater. Sci., 24 (1979), 143.
49) C.M. Wayman: Introduction to the Crystallography of Martensitic Transformations, New York, MacMillian, (1964).
50) J.W. Christian: Theory of Phase Transformations in Metals and Alloys, New York, Pergamon Press., (1965).
51) A. G. Khachaturyan: Theory of Structural Transformations in Solids, New York, Wiley, (1983).
52) H. Funakubo (ed.): Shape Memory Alloys, Gordon and Breach Science Publishers, (1987).
53) C. M. Wayman and K. Shimizu: Met. Sci. J., 6 (1972), 175.
54) L. Delaey, R. V. Krishnan, H. Tas and H. Warlimont: J. Mater. Sci., 9 (1974),1521.
55) K. Otsuka and C. M. Wayman: Deformation Behavior of Materials, ed. P. Feltham, Vol. II, Israel, Freund Publishing House, (1977), 98.
56) L. Delaey, M. Chandrasekaran, M. Andrade and J. Van Humbeeck: Proc. Int. Conf. Solid-solid Phase Transfomations, Pittsburgh, (1981), 1429.
57) K. Otsuka and K. Shimizu: Proc. ICOMAT-79, Cambridge, (1979), 607.
58) K. Otsuka and K. Shimizu: Proc. Int. Conf. of Solid-solid Phase Transformations, Pittsburgh, PA, (1987), 1267.
59) J. W. Christian: Met. Trans. A, 13 (1982), 509.
60) K. Otsuka and K. Shimizu: Met. Forum, 4 (1981), 125.
61) K. Otsuka and K. Shimizu: International Metals Reviews, 31 (1986), 93.
62) T. Tadaki, K. Otsuka and K. Shimizu: Ann. Rev. Mater. Sci., 18 (1988), 25.

63) K. Shimizu and T. Tadaki: Shape Memory Alloys, ed H. Funakubo, Gordon and Breach Science Publishers, (1987), 1.
64) M. S. Wechsler, D. S. Lieberman and T. A. Read: Trans. AIME, 197 (1953), 1503.
65) J. S. Bowles and J. K. Mackenzie: Acta Metall., 2 (1954),129,138,224.
66) D.S. Lieberman, M.S. Wechsler and T. A. Read: J. Appl. Phys., 26 (1955), 473.
67) K. Okamoto, S. Ichinose, K. Morii, K. Otsuka and K. Shimizu: Acta Met., 34 (1986), 2065.
68) S. Ichinose, Y. Funatsu and K. Otsuka: Acta Met., 33 (1985), 1613.
69) K. Otsuka and K. Shimizu: Scripta Met., 11 (1977), 757.
70) M. Nishida and T. Honma: Scripta Metall., 18 (1984), 1293.
71) D. P. Dautovich and G. R. Purdy: Can. Metall. Q., 4 (1965), 129.
72) S. Miyazaki, K. Otsuka and C. M. Wayman: this issue.
73) K. H. Eckelmeyer: Scripta Metall., 10 (1976), 667.
74) J. E. Hanlon, S. R. Butler and R. J. Wasilewski: Trans. TMS-AIME, 239 (1967), 1323.
75) W. B. Cross, A. H. Kariotis and F. J. Stimler: NASA CR-1433, September 1969.
76) F. E. Wang, B. F. DeSavage, W. J. Buehler and W. R. Hosler: J. Appl. Phys., 39 (1968), 2166.
77) G.P. Sandrock, A.J. Perkins and R.F. Hehemann: Metall. Trans., 2 (1971), 2769.
78) J. Perkins: Met. Trans., 4 (1973), 2709.
79) C. M. Wayman, I. Cornelis and K. Shimizu: Scripta Metall., 6 (1972), 115.
80) S. Miyazaki, Y. Igo and K. Otsuka: Acta Met., 34 (1986), 2045.
81) T. Tadaki, Y. Nakata and K. Shimizu: Trans. JIM, 28 (1987), 883.
82) H. Tamura, Y. Suzuki and T. Todoroki: Proc. ICOMAT-86, Nara, (1986), 736.
83) T. Todoroki, H. Tamura and Y. Suzuki: Proc. ICOMAT-86, Nara, (1986), 748.
84) G. B. Stachowiak and P. G. McCormick: Acta Metall., 36 (1988), 291.
85) G. Airoldi, G. Bellini and C. Di Francesco: J. Phys. F: Met. Phys., 14 (1984), 1983.
86) K. N. Melton and O. Mercier: Acta Met., 27 (1979), 137.
87) J.L. McNichols, Jr., P.C. Brookes and J. S. Cory: J. Appl. Phys., 52 (1981), 7442.
88) S. Miyazaki, Y. Sugaya and K. Otsuka: Proc. MRS Inter. Meeting on Advanced Materials, Tokyo, (1988), 251.
89) S. Miyazaki, Y. Sugaya and K. Otsuka: Proc. MRS Inter. Meeting on Advanced Materials, Tokyo, (1988), 257.
90) S. Miyazaki, M. Suizu, K. Otsuka and T. Takashima: Proc. MRS Inter. Meeting on Advanced Materials, Tokyo, (1988), 263.
91) S. Miyazaki, I. Shiota, K. Otsuka and H. Tamura: Proc. MRS Inter. Meeting on Advanced Materials, Tokyo, (1988), 153.
92) S. Miyazaki, A. Ishida and A. Takei: Proc. Intern. Symp. on Measurement and Control in Robotics (ISMCR-92), Tsukuba (1992), 495.
93) K. Nomura, S. Miyazaki and A. Takei: Trans. Mat. Res. Soc. Jpn., Vol. 18B (1994), 1049.
94) S. Miyazaki, S. Kurooka, A. Ishida and M. Nishida, Trans. Mat. Res. Soc. Jpn., Vol. 18B (1994), 1041.

95) S. Miyazaki and A. Ishida: Materials Transactions, JIM, (1994), 14.
96) P. Duval and P. Hayman: Memoires Scientifiques Rev. Metallurg., 70, No.2 (1973).
97) D.P.Dunne and N.F.Kennon : Metals Forum, 4 (1981),176.
98) M.Hansen : Constitution of Binary Alloys, McGrow-Hill Book Company, Inc, p.650 (1958).
99) L.Delaey, A.Deruyttere, N.Aernoudt and J.R.Ross : INCR Research Report (Project No.238), Febaruary, (1978).
100) S.Miyazaki, S.Ichinose and K.Otsuka : unpublished work
101) S.Miyazaki, K.Shibata and H.Fujita : Acta Met.,27 (1979),855.
102) I.Dvorak and E.B.Hawabolt : Met. Trans.A, 6 (1975),95.
103) F.Nakamura, J.Kusui, Y.Shimizu and J.Takamura : J.Japan Inst. of Metals, 44 (1980) 1302.
104) S.Miyazaki, K.Otsuka, H.Sakamoto and K.Shimizu : Trans. JIM, 22 (1981),244.
105) K.Tabei and M.Hatsujika : Abstracts of the Annual Spring meeting of the Japan Institute of Metals, (1982),105.
106) J.Perkins and W.E.Muesing : Met. Trans.A, 14 (1983),33.
107) T.Kikuchi and S.Kajiwara : Proc. 5th Int. Conf. on High Voltage Electron Microscopy, Kyoto, (1977),607.
108) Y.Nakata, T.Tadaki and K.Shimizu : Trans. JIM, 26(1985),646.
109) D.Schofield and A.P.Miodownik : Metals Technology, April,(1980),167.
110) G.Inden : Z.Metallkd.,66 (1975),648.
111) N.F.Kennon, D.P.Dunne and L.Middleton : Met. Trans. A, 13(1982),551.
112) G.Scarsbrook, J.Cook and W.M.Stobbs: Proc. ICOMAT-82, Belgium, (1982),p.C4-703.
113) H.Sakamoto, Y.Kijma and K.Shimizu : Trans. JIM, 23(1982),585.
114) H.Sakamoto : Trans. JIM, 24(1983),665.
115) S.Miyazaki, T.Kawai and K.Otsuka: Scripta Met., 16(1982),431.
116) S.Miyazaki, T.Kawai and K.Otsuka : Proc. ICOMAT-82,belgium, (1982);C4-85.
117) J.S.Lee and C.M.Wayman : Trans. JIM,27 (1986),584.
118) G.N.Sure and L.C.Brown : Met. Trans.A,15,(1984),1613.
119) Y.Ikai, K.Murakami and K.Mishima : Proc. ICOMAT-82, Belgium (1982),C4-785.
120) K.Enami, N.Takimoto and S.Nenno : Proc. ICOMAT-82, Belgium, (1982),C4-773.
121) Y.Hanadate, M.Miyagi, T.Hamada and F.Uratani: Osaka Industrial Engineering Research Center Repot, No.81,(1982),17.
122) K.Adachi, Y.Hamada and Y.Tagawa : Scripta Met.,21(1987),453.
123) K.Sugimoto, K.Kamei, H.Matsumoto, S.Domatsu, K.Akamatsu and T.Sugimoto : Proc. ICOMAT-82, Belgium, (1982),C4-761.
124) G.N.Sure and L.C.Brown : Scripta Met., 19 (1985),401.
125) R.Oshima, M.Tanimoto, T.Oka, F.E.Fujita, Y.Hanadate, T.Hamada and M.Miyagi : Proc. ICOMAT-82, Belgium, (1982),C4-749.
126) J.Janssen, F.Willems, B.Verelst, J.Maertens and L.Delaey: Proc. ICOMAT-82,(1982),C4-809.

127) T.Takashima : Kobe Steel Engng. Reports, 37(1987),40.
128) R.Oshima, S.Sugimoto, M.Suguyama, T.Hamada and F.E.Fujita :Trans. JIM, 26(1985),523.
129) C.M.Wayman : Scripta Met., 5(1971),489.
130) M.Foos, C.Frantz and M.Gantois : Acta Met.,29(1981),1091.
131) S.Muto, R.Oshima and F.E.Fujita : Proc. ICOMAT-86, Nara, (1986),997.
132) R.Oshima : Proc. ICOMAT-86,Nara,(1986),971.
133) M.Maki and I.Tamura : Bulletin of the Japan Institute of Metals, 23(1984),229. (in Japanese)
134) S.Kajiwara, T.Kikuchi and N.Sakuma : Proc. ICOMAT-86, Nara, (1986),991.
135) M.Watanabe and C.M.Wayman : Met. Trans., 2(1971),2221,2229.
136) T.Sanpei and Y.Moriya : private communication.
137) S. Miyazaki: Thermal and stress cycling effects and fatigue properties in Ti-Ni alloys, in: Engineering Aspects of Shape Memory Alloys (Ed. T.W.Duerig et al.), Butterworths (1990), 452
138) S. Miyazaki: Proc. IEEE Micro Electro Mechanical Systems (MEMS-94), Oiso, Japan, (1994), 176.
139) S. Miyazaki: Special Issue on Micromachines and Materials Science, Metals, Vol.63, No.3 (1993). (in Japanese)

Printed in the United States
By Bookmasters